THE UNITED STATES AND EUROPE

The contemporary relationship between America and Europe is both fraught and dynamic. Post-war reconstruction of Europe brought economic and political integration – and the creation of a 'United States of Europe' was a goal shared by many Americans. Yet the influence of neo-conservatism on American foreign policy and the contemporary 'War on Terror' has re-defined relationships between America and Europe, both 'old' and 'new'. Adopting an interdisciplinary approach to explore the historical, political, economic and cultural dimensions of the transatlantic relationship, this volume brings together experts from Britain, Europe and America to analyse a political relationship that remains fundamental to the maintenance of global security. Providing an in-depth analysis of the challenges that have been met and those that have to be faced, the editors have succeeded in producing an accessible and informative text that incorporates the latest research in the field.

This book will be of interest to undergraduate and postgraduate students of US foreign policy, International Relations and American and European politics.

John Baylis is Professor of Politics and International Relations and Pro-Vice-Chancellor at Swansea University. He is the author of over twenty books, including *Anglo-American Defence Relations* (1984) and *Anglo-American Relations since 1939: The Enduring Alliance* (1994).

Jon Roper is Professor of American Studies at Swansea University. He is the author of *Democracy and its Critics* (1989), *The American Presidents* (2000) and *The Contours of American Politics* (2002).

CONTEMPORARY SECURITY STUDIES

Daniel Ganser, *NATO's Secret Army: Operation Gladio and Terrorism in Western Europe*

Peter Kent Forster and Stephen J. Cimbala, *The US, NATO and Military Burden-Sharing*

Irina Isakova, *Russian Governance in the Twenty-First Century: Geo-Strategy, Geopolitics and New Governance*

Craig Gerrard, *The Foreign Office and Finland 1938–1940: Diplomatic Sideshow*

Isabelle Duyvesteyn and Jan Angstrom (eds), *Rethinking the Nature of War*

Brendan O'Shea, *Perception and Reality in the Modern Yugoslav Conflict: Myth, Falsehood and Deceit 1991–1995*

Tim Donais, *The Political Economy of Peacebuilding in Post-Dayton Bosnia*

Peter H. Merkl, *The Rift between America and Old Europe: The Distracted Eagle*

Jan Hallenberg and Håkan Karlsson (eds), *The Iraq War: European Perspectives on Politics, Strategy, and Operations*

Richard L. Russell, *Strategic Contest: Weapons Proliferation and War in the Greater Middle East*

David R. Willcox, *Propaganda, the Press and Conflict: The Gulf War and Kosovo*

Bertel Heurlin and Sten Rynning (eds), *Missile Defence: International, Regional and National Implications*

Chandra Lekha Sriram, *Globalising Justice for Mass Atrocities: A Revolution in Accountability*

Joseph L. Soeters, *Ethnic Conflict and Terrorism: The Origins and Dynamics of Civil Wars*

Brynjar Lia, *Globalisation and the Future of Terrorism: Patterns and Predictions*

Stephen J. Cimbala, *Nuclear Weapons and Strategy: The Evolution of American Nuclear Policy*

Owen L. Sirrs, *Nasser and the Missile Age in the Middle East*

Yee-Kuang Heng, *War as Risk Management: Strategy and Conflict in an Age of Globalised Risks*

Jurgen Altman, *Military Nanotechnology: Potential Applications and Preventive Arms Control*

Eric R. Terzuolog, *NATO and Weapons of Mass Destruction: Regional Alliance, Global Threats*

Pernille Rieker, *Europeanisation of National Security Identity: The EU and the Changing Security Identities of the Nordic States*

T. David Mason and James D. Meernik (eds), *International Conflict Prevention and Peace-building: Sustaining the Peace in Post Conflict Societies*

Brian Rappert, *Controlling the Weapons of War: Politics, Persuasion, and the Prohibition of Inhumanity*

Jan Hallenberg and Håkan Karlsson (eds), *Changing Transatlantic Security Relations: Do the US, the EU and Russia Form a New Strategic Triangle?*

Thomas M. Kane, *Theoretical Roots of US Foreign Policy: Machiavelli and American Unilateralism*

Christopher Kinsey, *Corporate Soldiers and International Security: The Rise of Private Military Companies*

Gordon Adams and Guy Ben-Ari, *Transforming European Militaries: Coalition Operations and the Technology Gap*

Robert G. Patman (ed.), *Globalisation, Conflict and the Security State: National Security in a 'New' Strategic Era*

James V. Arbuckle, *Military Forces in 21st Century Peace Operations: No Job for a Soldier?*

Nick Ritchie and Paul Rogers, *The Political Road to War with Iraq: Bush, 9/11 and the Drive to Overthrow Saddam*

Michael A. Innes (ed.), *Bosnian Security after Dayton: New Perspectives*

Andrew Priest, *Kennedy, Johnson and NATO: Britain, America and the Dynamics of Alliance, 1962–68*

Denise Garcia, *Small Arms and Security: New Emerging International Norms*

John Baylis and Jon Roper (eds), *The United States and Europe: Beyond the Neo-Conservative Divide?*

THE UNITED STATES AND EUROPE

Beyond the Neo-Conservative Divide?

Edited by John Baylis and Jon Roper

LONDON AND NEW YORK

First published 2006
by Routledge
2 Park Square, Milton Park, Abingdon, Oxon OX14 4RN

Simultaneously published in the USA and Canada
by Routledge
270 Madison Ave, New York, NY 10016

Routledge is an imprint of the Taylor & Francis Group
an informa business

Typeset in Times by BC Typesetting Ltd, Bristol
Printed and bound in Great Britain by
Antony Rowe Ltd, Chippenham, Wiltshire

British Library Cataloguing in Publication Data
A catalogue record for this book is available from the British Library

Library of Congress Cataloging in Publication Data
A catalog record has been requested for this book

ISBN10: 0–415–36829–4 (hbk)
ISBN10: 0–415–36999–1 (pbk)
ISBN10: 0–203–02757–4 (ebk)

ISBN13: 978–0–415–36829–2 (hbk)
ISBN13: 978–0–415–36999–2 (pbk)
ISBN13: 978–0–203–02757–8 (ebk)

This book is dedicated with love to Emma and Katie Baylis
and to Caitlin, Aisling and Jack Roper

CONTENTS

CONTRIBUTORS

John Baylis is Professor of Politics and International Relations and Pro-Vice-Chancellor at Swansea University. He is the author of over twenty books, including *Anglo-American Defence Relations 1939–1984* (1984), *The Diplomacy of Pragmatism* (1993), *Anglo-American Relations since 1939: The Enduring Alliance* (1997) and (edited with Steve Smith) *The Globalization of World Politics* (3rd edn, 2005).

Saki Ruth Dockrill is Professor of Contemporary History and International Security at the Department of War Studies, King's College, London, and is a former John M. Olin Fellow at Yale University. Her publications include *Eisenhower's New Look National Security Policy* (1996), *Britain's Retreat from East of Suez: The Choice between Europe and the World?* (2002) and *The End of the Cold War Era: The Transformation of the Global Security Order* (2005).

Beatrice Heuser is currently teaching at the University of the Bundeswehr in Munich. She was Professor of International and Strategic Studies in the Department of War Studies, King's College, London. Her publications include: *Nuclear Mentalities? Strategies and Beliefs in Britain, France, and the FRG* (1998) and (edited with Anja Hartmann) *War, Peace and World Orders in European History* (2001).

Walter W. Hölbling teaches US literature and culture at the University of Graz. Among his recent publications are *What Is American? New Identities in U.S. Culture* (2004) and *'Nature's Nation' Revisited: American Concepts of Nature from Wonder to Ecological Crisis* (2003), as well as essays on American Studies in Europe, US postcolonial fiction, and affinities between American literature and political rhetoric.

Jolyon Howorth has been Visiting Professor of Political Science at Yale since 2002. He is Jean Monnet Professor of European Politics *ad personam* at the University of Bath. His books include: (edited with John Keeler) *Defending Europe: The EU, NATO and the Quest for European Autonomy*

(2003), *European Integration and Defence: The Ultimate Challenge?* (2000) and *Defence Policy in the European Union* (2006).

Wade Jacoby is Associate Professor of Political Science and Director of the Center for the Study of Europe at Brigham Young University. His books include *Imitation and Politics: Redesigning Modern Germany* (2001) and *The Enlargement of the EU and Nato: Ordering from the Menu in Central Europe* (2004).

Kerry Longhurst is Senior Lecturer in European Security at the European Research Institute (ERI), University of Birmingham. Her research interests focus upon security issues in the 'wider' Europe. Her publications include *Germany and the Use of Force* (2004) and (edited with Marcin Zaborowski) *Old Europe, New Europe and the Transatlantic Security Agend*a (2005).

Joseph A. McKinney is Ben H. Williams Professor of International Economics at Baylor University. His major area of interest is international trade policy, particularly regional economic integration. He has published articles in professional journals and has authored or edited five books and monographs in this area, including: *Created from Nafta: The Structure, Function, and Significance of the Treaty's Related Institutions* (2000).

Steve Marsh is Lecturer in the School of European Studies at Cardiff University. His research interests lie in Anglo-American relations, US foreign policy and the external relations of the European Union. His publications include *Anglo-American Relations and Cold War Oil* (2003) and two co-authored books: *The International Relations of the European Union* (2005) and *US Foreign Policy Since 1945* (2nd edn, 2006).

Craig Phelan is Senior Lecturer in the Department of American Studies at Swansea University. He is editor of the journal *Labor History*, and author of numerous articles and books on that subject including the edited collection, *The Future of Organised Labour: Global Perspectives* (2006).

Jon Roper is Professor and Head of the Department of American Studies at Swansea University. His research interests encompass American political ideas, the American Presidency and the impact of war on American politics, culture and society. His publications include *Democracy and Its Critics* (1989), *The American Presidents: Heroic Leadership from Kennedy to Clinton* (2000) and *The Contours of American Politics* (2002).

Jean-marie Ruiz is Associate Professor of American Studies at the Université de Savoie. His research interests are in the history of American political thought, and its connections with European political thought. He is the co-editor of a book on current transatlantic relations: *Modèle ou Repoussoir? Les États-Unis et l'Europe Depuis la Fin de la Guerre Froide* (2006).

Stan A. Taylor is Emeritus Professor of Political Science at Brigham Young University, former Visiting Professor in the Department of American Studies at Swansea University, and Visiting Professor in the Political Studies Department at the University of Otago. His recent publications include: 'Intelligence Reform: Will More Agencies, Money, and Personnel Help?' *Intelligence and National Security* (2004) and 'Intelligence' in A. Collins (ed.) *Contemporary Security Studies* (2006).

ACKNOWLEDGEMENTS

The editors wish to express their thanks to Kristan Stoddart for his assistance with the arduous task of editing this volume.

ABBREVIATIONS AND ACRONYMS

AFSOUTH	Allied Forces South
BiH	Bosnia-Herzegovina
BRUS	Britain US Agreement
CAP	Common Agricultural Policy
CDU	Christian Democratic Union (of Germany)
CEE	Central and Eastern Europe
CEEC	Central and Eastern European Countries
CESDP	Common European Security and Defence Policy
CFSP	Common Foreign Security Policy
CJTF	Combined Joint Task Force
COPS	Political and Security Committee
CPCO	Centre de Planification et de Conduite
CSU	Christian Social Union (of Bavaria)
EADC	European Aerospace and Defence Company
ECAP	European Capability Action Plan
EDA	European Defence Agency
EDC	European Defence Community
ERP	European Recovery Programme
ESDI	European Security and Defence Identity
ESDP	European Security and Defence Policy
ESS	European Security Strategy
EU	European Union
EUIS	European Union Intelligence Service
EUMC	European Union Military Committee
EUMS	European Union Military Staff
FCO	Foreign and Commonwealth Office
FYROM	Former Yugoslav Republic of Macedonia
GDP	Gross Domestic Product
HG	Headline Goal
HR-CFSP	High Representative for the CFSP
HUMINT	Human Intelligence
ICC	International Criminal Court

ISAF	International Security Assistance Force
MICI	UK Wartime Foreign Intelligence
NAC	North Atlantic Council
NATO	North Atlantic Treaty Organisation
NGO	Non-governmental Organisation
NRF	NATO Response Force
NTA	New Transatlantic Agenda
NYSE	New York Stock Exchange
OECD	Organisation for European Economic Co-operation
OSCE	Organisation for Security and Co-operation in Europe
OSS	Office of Strategic Services
PfP	Partnership for Peace
PGM	Precision Guided Missiles
PNR	Passenger Name Record
PRO	Public Record Office
PRT	Provincial Reconstruction Team
PU	Policy Unit
R&D	Research and Development
SACEUR	Supreme Allied Commander in Europe
SFOR	Stabilisation Force
SHAPE	Supreme Headquarters Allied Powers Europe
SIGINT	Signals Intelligence
SOE	Special Operations Executive
SPD	Social Democratic Party (of Germany)
TCN	Troop Contributing Nation
TEU	Treaty on European Union
UKUS	United Kingdom, United States, Canada, Australia
UNSC	United Nations Security Council
UNSCOM	United Nations Special Commission
WEU	Western European Union
WMD	Weapons of Mass Destruction
WTO	World Trade Organisation

1

INTRODUCTION

Jon Roper and John Baylis

There is a deceptive symmetry in two dates that have shaped symbolically the contemporary political and military relationship between the United States and Europe. November 9 1989 and September 11 2001 encapsulate the twelve years between the fall of the Berlin Wall and the collapse of the twin towers of the World Trade Center in New York. In the United States, the ebullient optimism that accompanied the end of the Cold War was dramatized by the events in Berlin just as the shock of the terrorist attacks on its mainland forced Americans into an immediate re-evaluation of the threats which confront them in an uncertain world. In Europe, after the Wall was breached the challenges of further economic and political integration were juxtaposed with the resurgence of ethnic conflict in the Balkans. With its political map changing rapidly, maintaining peace and security within the newly defined borders of continental Europe became a new mission for a NATO alliance that had been formed in the crucible of Cold War containment. Moreover, less than twelve months after Berlin had become the transient focus of popular attention, Iraq invaded Kuwait.

On September 11 1990, President George Bush addressed a joint session of Congress and argued that

> the crisis in the Persian Gulf . . . offers a rare opportunity to move towards a historic period of cooperation. Out of these troubled times . . . a new world order – can emerge; a new era – freer from the threat of terror, stronger in the pursuit of justice, and more secure in the quest for peace, an era in which the nations of the world, East and West, North and South, can prosper and live in harmony.[1]

European nations joined the United States in the military coalition which, with the approval of the United Nations, forced Saddam Hussein into retreat. But if the historic opportunity that the President predicted was ever anything other than a rhetorical flourish, it was one which the subsequent course of events – not least in Iraq – would show was clearly missed.

Following the events that occurred on the eleventh anniversary of his father's speech to Congress, President George W. Bush thus saw the world in starkly different terms. In his address to the nation on September 20 2001 he was unequivocal:

> every nation, in every region, now has a decision to make. Either you are with us, or you are with the terrorists. From this day forward, any nation that continues to harbour or support terrorism will be regarded by the United States as a hostile regime.[2]

It was a prelude to American-led military action to overthrow the Taleban in Afghanistan, and then to Iraq becoming the focus of American hostility in the President's 'war on terror'. At that point, the transatlantic alliance between much of Europe and the United States was placed under such political stress that it fractured. Is it now beyond repair? Or has there been a reconfiguration of relationships between and among America and the nations that the President's secretary of defence, Donald Rumsfeld, characterized – and caricatured – as 'Old' and 'New' Europe?

The chapters in this book analyse the transatlantic relationship from a number of different perspectives to assess whether the political divide between the two continents that resulted from America's pre-emptive action in the Middle East is a permanent signpost to the future or a transient reflection of contemporary ideological perspectives. They look at the contexts in which the connections between America and Europe may be placed: historical, political, economic and cultural. At the outset, there is recognition that the contemporary debate over the future of Europe's relations with America has been informed by the ideological outlook of neo-conservatism in the United States and framed by the President's military responses to the events of September 11.

In his influential study *Paradise and Power: America and Europe in the New World Order* (2003), Robert Kagan argues that it is time to stop pretending that Europeans and Americans share a common view of the world – or even that they occupy the same world. On the all-important question of power – the efficacy of power, the morality of power, and the desirability of power – he suggests that American and European perspectives are very different. Europe is turning away from power – or rather moving beyond power into a self-contained world of laws and rules and transnational negotiation and cooperation. Kagan claims that Europe is entering a post-historical paradise of peace and relative prosperity – which reflects the realization of Kant's notion of 'perpetual peace'. In contrast he sees the United States 'mired in history, exercising power in an anarchic Hobbesian world where international laws and rules are unreliable, and where true security and the promotion of a liberal order still depend on the possession and use of military might'.[3] For Kagan, then, the provocative conclusion is that a planetary divide now defines transatlantic relations: whereas Americans are from Mars, Europeans are from Venus.

Kagan's overall view of the transatlantic relationship is very pessimistic. Americans and Europeans agree on little and understand one another less and less. He argues that this is not a transitory state of affairs – the product of one American election or one catastrophic event like 9/11 – but rather a permanent split. In his view:

> The reasons for the transatlantic divide are deep, long in development, and likely to endure. When it comes to settling national priorities, determining threats, defining challenges, and fashioning and implementing foreign and defence policies, the United States and Europe have parted ways.[4]

One of the central questions discussed in this book is how serious the transatlantic rift is. A case can certainly be made that Kagan over-exaggerates the Hobbesian and Kantian perspectives of the US and Europe – there are clearly some Hobbesian tendencies in European policy, as we have seen with Bosnia and Kosovo. He also ignores the differences within Europe itself – between Britain, Spain and Poland, and France, Belgium and Germany during the Iraq War. Nevertheless, there is little doubt about the serious nature of the rift which has appeared in transatlantic relations during recent years. There were numerous differences during the Cold War – over Suez, French withdrawal from NATO's military structure, and questions of burden-sharing – but there was a basic understanding of the shared threat, which prevented a permanent schism.

Since the end of the Cold War, as some of the realist writers predicted, the number of differences have grown over trade relations, over the Middle East, over European defence, the future of NATO, and over how to deal with 'rogue states'. Despite these disputes, however, there was still a sense of a 'Western identity' – of allies working together – even if they sometimes fell out. That sense of identity was very much on show in the immediate aftermath of 9/11. After the attacks on New York and Washington what seems to have changed is the new focus in the US on counter-terrorism, the related anxieties about weapons of mass destruction and the move towards preemptive strategies. The new primacy given to counter-terrorism by the Bush administration and a greater emphasis on unilateralism, together with active opposition from some European states, have led to new tensions in the transatlantic relationship which threaten to undermine the common identity in a way that has never been the case in the past.

Kagan's polemic represents one of the best-publicized neo-conservative arguments for the permanence of a transatlantic divide.[5] There are, however, other neo-conservative accounts of the difficulties in contemporary European–American relations which focus less on differing views about power and more on the spiritual divide. In his book *The Cube and the Cathedral: Europe, America and Politics Without God* (2005),[6] George Weigel argues that important differences over human rights and democracy on both sides of the Atlantic are fundamentally the result of the atheistic humanism of nineteenth-century European intellectual life which continues to pervade European life today. It is only societies like the United States ('the Cathedral') that can show others their commitment to freedom. In contrast, the Europeans (the 'people of the Cube'), Weigel argues, are incapable of providing a convincing moral argument in defence of human rights and democracy, because they have lost sight of God.

It is the challenge of neo-conservative approaches to the transatlantic relationship which forms the focus for this study. In his contribution to this book, Jon Roper (Chapter 2) argues from a historical perspective that this distinction between America and Europe is neither as stark nor as enduring as the neo-conservatives believe. At the heart of the neo-conservative persuasion is a desire to project Presidential – and American – power through successful military intervention abroad. On the other hand, since the Second World War, many Europeans have seen dangers in attempting to find military solutions to political problems, a lesson that some Americans too derived from their experience of the Vietnam War. Americans who are not persuaded by the neo-conservative case thus share the European outlook that was expressed most visibly in opposition to George W. Bush's military action against Iraq. Moreover, the challenges overcome by Americans, in designing their 'novus ordo seclorum' (a new order of the ages), and in writing the United States Constitution in 1787, are similar to those now faced by those who advocate Europe's economic and political integration as a means of maintaining peace within its borders – however elastically they may be defined. The United States itself thus provides Europeans with a historical example of reaching for Venus rather than settling for Mars in suggesting how state sovereignties may be merged in the interests of greater security.

In Chapter 3, Kerry Longhurst discusses the claim that America's transatlantic relationships can be framed in the context of 'Old' and 'New' Europe and assesses whether there has indeed been a 'recalibration' of transatlantic security relations in the post-9/11 period. Taking Germany and Poland as representative examples of the old and the new, differing attitudes to contemporary American foreign policy – particularly with respect to Iraq – may be seen as a product of distinctive national histories and responses to the end of the Cold War. To the extent that such differences exist and persist among the states of an enlarged European Union, the outcome will be a transatlantic alliance that is less cohesive internally and in which strategic priorities may continue to differ among its members. Moreover, the idea that European nations may be divided simply into 'Europeanist' or 'Atlanticist' camps, or even into the 'Old' and the 'New', ignores the complexities that frame the continuing evolution of transatlantic relationships.

The critique of neo-conservatism is taken up by Jean-marie Ruiz in Chapter 4. In discussing responses to Robert Kagan's polemic in France, Ruiz points out that among those who have been most trenchant in their criticisms have been Jean-Marc Ferry and Pierre Hassner, both of whom support the ideal of European political and economic integration. Moreover, for Ferry the concept of power in international relations may be expressed in different ways. Europeans do not make an explicit connection between political influence and military power. International prestige may not be directly related to the ability to fight wars. Indeed, what neo-conservatives admire as a source of strength – the capacity to project military power abroad – may engender resistance rather than respect, undermining American prestige and its long-run ability to dominate world affairs. Furthermore, as Hassner argues, the assumption of a fundamental divergence in European and American perspectives on the way to maintain international security and order is an over-simplification. In this sense, Kagan's thesis, provocative as it is, becomes a caricature of a more complex debate that persists on both sides of the Atlantic as well as between Europe and the United States.

In Chapter 5, Wade Jacoby focuses on US–German relations, and examines how political parties in Germany have reacted to domestic popular opinion as it has shifted away from initial support for America in the immediate aftermath of 9/11 towards more outright opposition to the perceived neo-conservative agenda of pre-emptive military action in the 'war on terror'. Although Gerhard Schröder initially managed to accrue electoral capital from this sentiment, his failure to deal with Germany's economic problems contributed to his narrow defeat in the elections of September 2005. However, the new Chancellor, Angela Merkel, is inexperienced on the international stage, and the lack of a clear mandate for the new German government suggests that in the immediate future the decisions that she may have to make in relation to America's actions abroad may be as hard as those which faced her predecessor.

For Tony Blair, however, the post-September 11 period brought a clarity to the 'special relationship' between Britain and the United States even as his support for the military actions taken by George W. Bush steadily eroded his political standing within his own party, within the country, and within Europe. As John Baylis points out (Chapter 6), although Britain may have a conception of itself as bridging the Atlantic divide between the United States and continental Europe, the history of the post-war 'special relationship' demonstrates that at times of international crisis the nation places itself firmly on the American side of that bridge. If George W. Bush's actions have fractured his transatlantic relationships with some European leaders, the British government's support for America's military involvement in Iraq has dealt a blow to its expressed ambition to remain at the heart of Europe. Maintaining the historic importance of the 'special relationship' while strengthening Britain's ties with its continental neighbours remains a delicate balancing act for those who seek to shape the future of British foreign policy. The bridge indeed may sometimes appear to be a tightrope.

Whatever the future of transatlantic relations, however, defence, security and intelligence will remain issues of common concern, and, ideological differences aside, will continue to encourage pragmatic cooperation between Europe and America. In Chapter 7, Steve Marsh examines American approaches to the development of a European security and defence identity involving efforts to forge a common policy in this area. The United States may see advantages in Europe looking towards Mars rather than Venus in terms of sharing the burdens of international peace-keeping, although this is necessarily seen within the context of the continuing importance to America of the NATO alliance. The United States still feels the need to be involved in Europe's efforts to build an autonomous strategic security and defence policy, to try to influence the shape of things to come: particularly to the extent that Europe sees an international role for itself outside the framework of NATO. Jolyon Howorth (Chapter 8) focuses on the future of NATO, considered by some to be threatened by the movement to establish a coordinated European foreign and defence policy. Indeed, there is the further issue of whether Europe could, or should, develop a military capacity that would allow it to intervene in international disputes beyond its own borders: a question that lies at the heart of the debate as to how NATO can accommodate itself to the realities of the contemporary world.

During the Cold War, NATO symbolized the desire of the United States and its European allies to cooperate militarily in the face of the perceived threat from an expansionist, communist, Soviet Union. As Saki Dockrill points out in Chapter 9, the events of September 11 2001 forced America and Europe to pay renewed attention to global security issues beyond their own borders, but their willingness to act in concert ended in the disharmony that accompanied American-led action against Iraq. Yet it is still an open question whether this indicates a 'paradigm shift' in transatlantic relations, shaped by a combination of historical, cultural and structural trends, as well as by the conflicting ideological outlooks now espoused by elites on either side of the Atlantic. If the fracturing of the alliance is irreparable, at the least it will result in the priorities of Europe becoming a peripheral influence and concern in the formation of American foreign policy.

On the other hand, Stan Taylor argues (Chapter 10) that one key to preserving domestic security in the face of terrorist threats is the cooperation of intelligence agencies across national boundaries. In the aftermath of the Second World War, the informal network of intelligence-sharing among the agencies of the Anglo-American democracies (the United States, Canada, the United Kingdom, Australia and New Zealand) offered an example which, apart from limited cases within NATO, was not replicated within the emerging European Union. Nevertheless, it is 'the spies that bind'. Whatever the political impact of turbulence within the transatlantic alliance, the pragmatic imperatives of intelligence-gathering encourage both formal and informal relationships to be developed and to be maintained among those charged with the responsibility of gathering the intelligence that can disrupt potential terrorist attacks.

Similarly, transatlantic cooperation is a necessity within the global economy. In Chapter 11, Joseph McKinney examines the level of interdependence that exists between the United States and Europe, the two largest trading blocs in the world. From an economic perspective, the relationship between Europe and America remains strong, despite the contemporary political climate in which it is framed. Moreover, the transatlantic economy is at the heart of the broader global economy: agreements that open up world trade cannot be made without the support of transatlantic trading partners. While political federation at a global level remains a utopian thought, globalization may imply the development of forms of economic federalism, built upon the foundation of the cooperative enterprise that continues to characterize transatlantic economic relations.

Nevertheless, as Craig Phelan observes (Chapter 12), there remains a perceived disparity between America's attitude towards the construction of a welfare state – despite the impact of the New Deal and the Great Society – and the commitment of many European governments to the maintenance of state welfare provision in terms of benefits, education and health. This view can be challenged. There are very real differences in the welfare policies that exist in Europe, and this diversity persists despite fears that globalization has prompted a 'race to the bottom' that will flatten national differences and undermine welfare provision worldwide.

If a prevailing view that there are profound differences in attitudes towards welfare provision on either side of the Atlantic is misplaced, to what extent are transatlantic relations framed too in the context of cultural stereotypes that are useful merely as rhetorical and ideological ammunition in contemporary arguments on both sides of the Atlantic? For Beatrice Heuser (Chapter 13), Donald Rumsfeld's suggestion that there is now an 'Old' and a 'New' Europe is just one example of such cultural misunderstanding. American stereotypes of France and the UK can be deconstructed with reference to Hollywood's adaptations of narratives such as Mark Twain's *A Connecticut Yankee at the Court of King Arthur*, as well as its renditions of staples of British folklore (Robin Hood) and French literature (*Les Miserables* and *The Three Musketeers*). What emerges as prevalent in American popular culture is a nostalgic vision of Britain as 'Old Europe' jostling with an image of France as a harbinger – like America – of the future and the 'New': a reversal of the perception of Europe contained in Rumsfeld's politically charged comment.

Twain's satirical novel, written at the end of the nineteenth century, and recounting the adventures of a time-travelling American transplanted to medieval Europe, is also a reference point in Walter Hölbling's discussion of the shaping of the dominant discourse in American culture which has helped to define contemporary American perspectives on the wider world (Chapter 14). Twain, who doubted his generation's enthusiasm for American imperialism, shows how intervention abroad can spark unintended consequences. Yet the power of the ideas that have propelled the desire to expand the sphere of American influence – from a cluster of states clinging to the eastern seaboard, across the continent to the Pacific Ocean, to the acquisition of territories overseas, and now to a position of international pre-eminence – is evident throughout the history of the United States. The Puritan 'errand into the wilderness', 'the last best hope for mankind', 'manifest destiny', 'providential mission', 'a world safe for democracy', 'a new world order' are expressions of a faith that is both religious and secular. They have become part of the cultural baggage of the nation's leaders as America engages in its contemporary 'war on terror'.

How America and Europe negotiate their future relationship depends on the extent to which the two continents can accommodate their contemporary political and ideological priorities within the context of their historical, cultural and economic connections across the Atlantic divide. What is clear from many of these chapters, however, is that, despite the perceptions of enduring ties, the United States and the nations of an expanded European Union, severally or collectively, may look back on the time between November 9 1989 and September 11 2001 as the prologue to an uncertain future.

Notes

1 George Bush, 'Address before a Joint Session of the Congress on the Persian Gulf Crisis and the Federal Budget Deficit', September 11 1990. George Bush Digital Library: http://bushlibrary.tamu.edu/
2 Quoted in David Frum, *The Right Man* (London: Weidenfeld & Nicolson, 2003), p. 146.
3 Robert Kagan, *Paradise and Power: America and Europe in the New World Order* (London: Atlantic Books, 2003), p. 3.
4 Ibid., p. 4.
5 See Gary Dorrien, *Imperial Designs: Neoconservatism and the New Pax Americana* (London: Routledge, 2004). This book describes how the ideology of American global pre-eminence originated during the presidency of George H.W. Bush, developed in the 1990s, gained power with the election of George W. Bush, and reshaped American foreign policy after September 11 2001.
6 George Weigel, *The Cube and the Cathedral: Europe, America and Politics Without God* (New York: Basic Books, 2005). More recently, however, other prominent neo-conservatives have re-appraised their ideological position. See, for example, Francis Fukuyama, *America at the Crossroads: Democracy, Power and the Neo-conservative Legacy* (New Haven: Yale University Press, 2006).

2

THE IDEA OF AMERICA AND THE IDEA OF EUROPE

Presidential power and constitutional authority

Jon Roper

At the end of the eighteenth century, in the interval between the American and the French Revolutions, 'on major strategic and international questions' of the day, to play with Robert Kagan's catchy planetary metaphor, Americans were from Venus and Europeans were from Mars.[1] In France, the pacifist and *philosophe* Marquis de Condorcet reflected a contemporary view when he argued that

> Americans will . . . help to maintain peace in Europe by the force of their example. . . . Any idea of aggressive war undertaken for aggrandizement or conquest is condemned by the calm judgment of a humane and peaceable people. . . . What can the militaristic prejudices of Europe offer in rebuttal to this example?[2]

Condorcet glimpsed in the United States the prospect of a new world order in which supra-national institutions might adjudicate on international disputes before nations resorted to war.

Yet, as Kagan also observed in *Paradise and Power* (2003), 'the ambition to play a grand role on the world stage is deeply rooted in the American character',[3] and as European military power diminished so the ability of the United States to act and to dominate international relations increased. While Europeans journeyed from Mars towards Venus, the American trajectory proved, despite Condorcet's hope, to be in the opposite direction. The contemporary transatlantic relationship is framed in very different terms from those that mapped its course in the 1780s. The United States has come to represent force rather than example.

Condorcet was writing in response to a competition organised by the Academy of Lyons and sponsored by the Abbé Raynal, who had proposed a prize for the best essay on an important contemporary issue. The title he chose proved more memorable than the original entries to the contest – no winner was ever declared: 'Was the discovery of America a blessing or a curse to mankind? If it was a blessing, by what means are we to conserve and enhance its benefits? If it was a curse, by what means are we to repair the damage?'[4] For contemporary Americans to whom the answers to these questions were so self-evident that they were beyond argument, this was a none too subtle blending of cultural prejudice and arrogance, and an example of European anti-Americanism that soon became a familiar reference point in a perennial debate about the impact of the United States not only upon Europe but upon the wider world.

Indeed, for European nations – France among them – in the aftermath of September 11 2001, the Abbé's competition now might be institutionalised as an annual event. Before the end of that year, the United States had embarked upon a war in Afghanistan with support among some European powers. This was followed swiftly by the build-up to America's military action in Iraq. Transatlantic relations fractured. In 2002, nine of the fifteen countries in the European Union were opposed to the prospect of war in Iraq. For the American defence secretary, Donald Rumsfeld, France and Germany, the two nations most critical of the United States, now represented an 'old Europe' that was reluctant to face contemporary realities. In contrast, the ten candidate countries then applying for admission to an enlarged European Union – the accession treaties were signed in April 2003 – supported America's action. Indeed, on May 1 that year, the same day that George W. Bush proclaimed, on the deck of the *USS Lincoln*, that 'major combat operations' were at an end in Iraq, those countries became part of the European project. It meant that a majority of the Union might now be claimed as members of the 'coalition of the willing' (even if it was still a minority prepared to be involved militarily). For many Europeans, however, America's pre-emptive military action had been taken on a faulty pretext and without the support of the international community expressed through the United Nations.

Their opposition led to the counter-charge made by neo-conservatives in America, among them Kagan, who argued that it was Europe's relative military impotence that led it to prefer international diplomacy in comparison to the United States which, since it now possessed military power, should be prepared to use it in an increasingly dangerous world. His argument rested upon a critical assumption: that Americans accept that they are from Mars, and that there are no effective domestic constraints on the President acting as commander-in-chief. This proposition is more prescriptive than it is descriptive. Neo-conservatives are convinced that the American President should be able to use military power, because they believe that for the past thirty years his capacity to do so, limited by the aftershocks of America's defeat in Vietnam, has made the nation vulnerable to a growing terrorist threat. The 'Vietnam Syndrome' impacted upon the chief executive's willingness to use military force overseas without the sustained support of the American public. So war – in Afghanistan and in Iraq – becomes a neo-conservative test of whether the President has re-captured the political initiative in the exercise of military power as commander-in-chief.

Throughout the twentieth century, the relationship between Europe and America can also be defined in the context of war. The two world wars that resulted from European rivalries, and in which the United States played a pivotal role, were crucial signposts on their respective journeys between Venus and Mars. The Cold War was a recognition of the ideological chasm that existed between the 'idea of America' as expressed by Thomas Jefferson in the Declaration of Independence in 1776 and the manifesto of Karl Marx, published in Europe seventy-six years later. Moreover, in the aftermath of the fall of the Berlin Wall and the collapse of the Soviet Union, a new existential conflict – the 'war on terror' – shaped contemporary transatlantic relations. Yet given the American President's current propensity to commit military forces overseas in pre-emptive action, the executive's powers as commander-in-chief remain as controversial to Kagan's European and American critics as they were to Condorcet's contemporaries, at a time when he was not alone in hoping that Venus might prove a more permanent and appealing destination for the President than Mars.

Presidents, power and war

On Friday August 17 1787, those meeting at the Constitutional Convention in Philadelphia discussed the issue of war. The debate was sparked by the clause in the draft of the Constitution that empowered the Congress of the United States to 'make war'. Charles Pinkney, from South Carolina, was opposed to the idea of giving this power to the legislature as a whole, arguing that it should instead be confined to the Senate. Others disagreed. Pierce Butler, Pinkney's fellow delegate from South Carolina, thought that the President should have the war-making power, and could be trusted not to use it unless there was popular support for such action. But if Butler was prescient in seeing public opinion as a critical element in determining the outcome of presidential wars, his argument did not convince the Convention.

James Madison and Elbridge Gerry then proposed a compromise. Instead of giving Congress the power to make war, it should be able to 'declare war; leaving to the Executive the power to repel sudden attacks'. Gerry commented that he 'never expected to hear in a republic a motion to empower the Executive alone to declare war'. George Mason, from Virginia, summed up the debate. He was opposed to

> giving the power of war to the Executive, because not safely to be trusted with it; or to the Senate, because not so constructed as to

> be entitled to it. He was for clogging rather than facilitating war, but for facilitating peace. He preferred '*declare*' to '*make*'.

So the motion was passed, and Article One, section eight of the Constitution gave Congress the power to declare war. It was also agreed that the Legislature should be able 'to raise and support Armies, but no Appropriation of Money to that Use shall be for a longer Term than two years' and that Congress should 'provide and maintain a Navy'.[5]

The Founders' intent was clear. Evidently, given their suspicion of unchecked Executive power, they were not about to allow the President unilaterally to declare war. They were also unwilling to agree to the establishment and the funding of a standing army, particularly if the President was to be its commander-in-chief. Indeed, ten days after the debate over who should have the power to declare war, the Convention amended, without substantive debate, Article Two, section two of the Constitution. To the opening clause, 'The President shall be Commander in Chief of the Army and Navy of the United States, and of the Militia of the several States', were added the words, 'when called into the actual service of the United States'.[6] So when the Legislature declared war and raised an army, the Executive could command it, but the President would have no standing army at his disposal.

The Founders hoped to institutionalise checks and balances upon the Executive's war-making powers. By giving Congress the constitutional authority to declare war, they sought to move away from the European model by which the Executive – usually a monarch – could decide to commit the nation to conflict. As Thomas Jefferson wrote approvingly to James Madison: 'We have already given . . . one effectual check to the dog of war, by transferring the power of letting him loose from the Executive to the Legislative body, from those who are to spend to those who are to pay'.[7] In his debate with Alexander Hamilton in 1793 over President Washington's proclamation of neutrality in the war that had broken out between France and other European powers, and writing under the pseudonym Helvidius (the classical philosopher who thought the Roman Emperor's power should be checked by the Senate), Madison himself argued that:

> War is in fact the true nurse of executive aggrandizement. In war a physical force is to be created, and it is the executive will which is to direct it. In war the public treasures are to be unlocked, and it is the executive hand which is to dispense them. In war the honours and emoluments of office are to be multiplied; and it is the executive patronage under which they are to be enjoyed. It is in war, finally, that laurels are to be gathered, and it is the executive brow they are to encircle. The strongest passions and most dangerous weak-

> nesses of the human breast, ambition, avarice, vanity, the honourable or venial love of fame, are all in conspiracy against the desire and duty of peace. Hence it has grown into an axiom that the executive is the department of power most distinguished by its propensity to war: hence it is the practice of all states, in proportion as they are free, to disarm this propensity of its influence.[8]

During the nineteenth and twentieth centuries and now beyond, however, succeeding generations of Americans saw Presidents involve their country in conflict. The United States has committed military forces overseas on at least fifteen separate occasions. Yet Congress has declared war only five times. The last time was over sixty years ago – the Second World War. Since then, Presidents – both Democrat and Republican – have become commanders-in-chief of a standing army, navy and air force. They have fought six major wars. It is a self-evident truth, therefore, that whatever the Founders' intent, the President now has the resources at his disposal, and has assumed the power, to make war, even if constitutionally he still cannot declare it.

Why is this? What are the precedents that have eroded Congressional control over the use of America's military establishment? From John Adams presiding over the naval war with France, to Thomas Jefferson and then James Madison taking action against the Barbary pirate states of North Africa (as Presidents both tempering their political philosophies with pragmatism), to Theodore Roosevelt's involvement in conflict with Mexico, examples of unilateral executive action in fighting the nation's enemies in defence of its perceived self-interest are easy to find. Couple these with Lincoln's extension of Executive powers during the Civil War, Franklin Roosevelt's political and military manoeuvrings prior to official American involvement in the Second World War, Harry Truman in Korea and Lyndon Johnson and Richard Nixon in Vietnam and it is apparent that the balance of power between Executive and Congress with respect to the use of military force has shifted decisively even within the elastic boundaries of the Constitution.

Congress's power to declare war was a Constitutional right designed as a check and a balance on the actions of the nation's commander-in-chief. Yet political realities, technological innovation and the American way of war have all but rendered it obsolete. In times of national crisis, the President is expected to act and the Congress finds it easier to defer to than to oppose the Executive's tactical imperatives of the moment. Since the development of atomic weapons, it is the President who has the choice of whether and when to deploy the nation's nuclear arsenal. Moreover, the creation of the most formidable, lavishly equipped and largest military establishment of any country in the world, and the capacity to deploy it anywhere on the planet, gives the modern Chief Executive military power that would have inspired concern rather than envy among the Founders, and notably in the General who became the nation's first President.

At the same time, however, George Washington believed then, as George W. Bush has become convinced now, that the United States is vulnerable to attacks from abroad. And fears over national security, over the nation's very survival, as they persist over time, have become the existential force moulding American attitudes to the wider world and encouraging the President's journey to Mars. Historically such anxieties underpin the strategic context that has undermined Congress's involvement in matters of war. Fear has allowed the President as commander-in-chief to assume centre stage in determining the nation's military priorities.

It was in the latter half of the twentieth century that the existential fear about national security became intertwined with an existential war. The Cold War was never declared, but was all-consuming. It is no accident that it was in its shadow that the so-called 'Imperial Presidency' began to flourish. The fear of an un-American ideology that might subvert American values, culture and society produced McCarthyism at home and support for unprecedented investment in military capacity. The Department of Defense took over from the Department of War. Fear underlined Truman's Doctrine and Eisenhower's theory that countries would fall to communism like collapsing dominoes. This was the context in which Kennedy and then Johnson and Nixon confronted the perceived threat of communism in Southeast Asia. The Vietnam War was the biggest executive-driven foreign policy débâcle of the twentieth century. It was the war that America lost, and the Imperial Presidency crashed and burned in its flames and those of the Watergate scandal. What Jefferson, Madison and their contemporaries could not have predicted, then, were the consequences for the President's power if, as commander-in-chief, he committed the United States to a military adventure that failed. It would take fully two centuries for that concern to become a relevant issue in American politics: given substance as the last American helicopters clattered away from its embassy in Saigon in 1975.

The war in Southeast Asia gave rise to its eponymous syndrome: a reluctance of the American people to support similar interventions around the world. Those who learned this 'lesson of Vietnam' came to a similar conclusion to that finally reached by Europeans after the Second World War. War was something to be avoided rather than to be embraced as a way of advancing the national interest. The mood of the moment reinforced this belief. At the end of the Cold War, it seemed that western civilisation, defined as the dominance of liberal democracy and capitalism as organising political and economic principles, might lead to a New World Order in which the threat of war was diminished.

Indeed, that was the essence of Francis Fukuyama's thesis in the article he published in *The National Interest* in the summer of 1989, a few months prior to the collapse of the Berlin Wall. In 'The End of History', he argued from a Hegelian perspective that it should be recognised that

> the basic principles of the liberal democratic state could not be improved upon. The two world wars in this century and their attendant revolutions and upheavals simply had the effect of extending those principles spatially, such that the various provinces of human civilization were brought up to the level of its most advanced outposts, and of forcing those societies in Europe and North America at the vanguard of civilization to implement their liberalism more fully.

With the collapse of any viable alternatives to the liberal democratic state, progress might be made towards a reduction in international tensions. For Fukuyama, the 'end of history' did

> not by any means imply the end of international conflict per se. For the world at that point would be divided between a part that was historical and a part that was post-historical. Conflict between states still in history, and between those states and those at the end of history, would still be possible. There would still be a high and perhaps rising level of ethnic and nationalist violence, since those are impulses incompletely played out, even in parts of the post-historical world. . . . This implies that terrorism and wars of national liberation will continue to be an important item on the international agenda. But large-scale conflict must involve large states still caught in the grip of history, and they are what appear to be passing from the scene.

Nevertheless, writing as an American and tempted by the thought of the excitement of Mars, Fukuyama concluded by confessing that

> even though I recognize its inevitability, I have the most ambivalent feelings for the civilization that has been created in Europe since 1945, with its north Atlantic and Asian offshoots. Perhaps this very prospect of centuries of boredom at the end of history will serve to get history started once again.[9]

On September 11 2001 it did so with a vengeance.

In the aftermath of 9/11 American perspectives changed. The idea that America was among those nations that had reached the 'End of History', and was waiting for others to catch up, was superseded by a different neo-conservative argument that directly impacted upon transatlantic relations. Kagan, in *Paradise and Power*, forsook Hegel for other European philosophers, and argued that Europe was 'entering a post-historical paradise of peace and relative prosperity, the realization of Immanuel Kant's "perpetual peace"'. On the other hand, 'the United States remains mired in history, exercising power in an anarchic Hobbesian world where true security and the defense and promotion of a liberal order still depend on the possession and use of military might'.[10] Whereas Europe now preferred to solve international disputes by relying on supra-national institutions such as the United Nations, and to use diplomacy to achieve such goals, the United States accepted the necessity of using military power to solve political problems. Kagan's argument, and indeed the dominant neo-conservative view, thus revolved around the use of power defined in terms of the rehabilitation of the President's ability to commit American military forces overseas free from the constraint implied by the Vietnam syndrome. Moreover, to the extent that America had proven reluctant to use military power in the post-Vietnam period, neo-conservatives argued that the nation had shown itself vulnerable to attacks from those whose political culture and ideologies were still trapped in history.

The Gulf War of 1991, which America fought with European support, had promised to resolve this problem of presidential power in the post-Vietnam period. But it did not. In the moment of victory, for President George Bush, America's success had exorcised the memory of defeat. He seemed exultant: 'By God, we've kicked the Vietnam Syndrome once and for all.' But it soon appeared that the President was a victim of his own rhetoric. Six weeks later, in justifying his reluctance to overthrow Saddam Hussein, the President was less ebullient:

> all along I have said that the United States is not going to intervene militarily in Iraq's internal affairs and risk being drawn into a Vietnam-style quagmire. This remains the case. Nor will we become an occupying power with US troops patrolling the streets of Baghdad.[11]

For the neo-conservative commentator Charles Krauthammer, writing in the *Washington Post*, this was proof that the President himself was now the 'chief purveyor' of the Vietnam syndrome. In his actions after Iraq's defeat, 'he simply raised the specter of Vietnam, an analogy without substance, and let its signal power, the power of fear and defeatism, do the rest. Bush did not just prove that the Vietnam syndrome lives. He gave it new life.'[12]

For neo-conservatives, then, America's reluctance to go to war in the aftermath of Vietnam ran counter to their belief that military power was an essential bulwark of American influence in the world, and that the President should be in a position to exercise his responsibilities as commander-in-chief against perceived threats to national security. For them, the events of

September 11 2001, when the architectural symbols of America's military industrial complex – its economic base in New York City, its military headquarters and the focus of its political power in Washington DC – were destroyed, damaged and threatened, were a defining moment for the first presidential administration of the twenty-first century.

On June 22 2005, Karl Rove, in a speech to a conservative gathering in Manhattan, caused controversy by claiming that whereas 'Conservatives saw the savagery of 9/11 in the attacks and prepared for war; liberals saw the savagery of the 9/11 attacks and wanted to prepare indictments and offer therapy and understanding for our attackers'.[13] His remarks demonstrate the chasm that separates neo-conservatives in America from those whom they perceive as their ideological opponents: it is a divide based upon a fundamentally different understanding of the constitutional limits on presidential power and the efficacy of war. If defeat in Vietnam brought widespread popular rejection of the idea that military force can solve political problems, then Americans who adhere to that persuasion have something in common with those Europeans who took on the task of reconstructing their continent in the aftermath of the Second World War, and who continue to grapple with the problems of political integration in the wake of the Cold War. So it is not the case that all Americans have reached Mars, just as not all Europeans prefer Venus. Rather some Americans – neo-conservatives – in their attitudes to presidential power and the use of military force, are currently on a different ideological planet from many in their own country and in Europe.

Kagan, however, insists that it is possible to generalise across the transatlantic divide. In his view, moreover:

> Europeans, because of their unique historical experience of the past century – culminating in the creation of the European Union – have developed a set of ideals and principles regarding the utility and morality of power different from the ideals and principles of Americans, who have not shared that experience.[14]

But is it possible for Europe to reach consensus on a worldview, different from that of the United States and expressed through a common framework of political institutions in which national sovereignties are submerged? It is another contemporary irony that just as Europeans seem to have embraced Condorcet's hope that supra-national institutions such as the United Nations can solve international problems, so popular sentiment in Europe has begun to run against the prospect of further political integration as outlined in the proposed, but now effectively disposed, European Constitution.

The treaty establishing a Constitution for Europe, agreed in October 2004, envisaged a President of the European Council – the Heads of Government of the member states – who would not hold national office and who would be elected for a term of two and a half years, renewable once. The President, under article 1-22 of the Constitution, would, 'in that capacity, ensure the external representation of the Union on issues concerning its common foreign and security policy', but the limited role given to the office reflects a concern with constraining potential centralised authority.[15] Nevertheless, even this 'idea of Europe' failed to gather popular support in favour of ratification in France and in Denmark. So does Europe still have faith in Kant's vision? Is its journey, in Kagan's words, 'beyond power into a self-contained world of laws and rules and transnational negotiation and cooperation' still on course and can its member states accept Europe's constitutional authority to express an 'idea of Europe' and a common purpose?[16] Once again, Condorcet's contemporaries in America provide some useful historical insights relevant to this debate.

Constitutions and political culture

If the problem of presidential power can be framed in terms of American attitudes towards war, the desire to mould a United States of Europe is the product of an ambition to maintain what had hitherto been a fragile history of peace. Consider, then, the political contexts in which the United States was established and in which the project for greater Western European integration was initially framed. The move towards Union in America in the immediate post-independence period was motivated, again, by existential fear. There were different components to this: at one level, the fear was of domestic political turmoil if the newly independent states proved themselves incapable of self-government. Incidents like Shays' rebellion in Massachusetts were seen as significant straws in the wind. Among those who gathered in Philadelphia in 1787 were those who, like James Madison, believed that their convention would 'decide forever the fate of republican government' in America.[17] But beyond this there was a greater fear, which is evident from even a cursory reading of the *Federalist*, written by Madison, Hamilton and John Jay as part of the effort to persuade the states – notably New York – to ratify the convention's work. This was that if America did not come together, not only would it fall apart, but also it would resemble Europe itself. As Hamilton observed: 'to look for a continuation of harmony between a number of independent, unconnected sovereignties in the same neighborhood, would be to disregard the uniform course of human events, and to set at defiance the accumulated experience of ages'.[18] In other words, if the American states did not agree to Union, then Europe's experience was a warning. The journey from Venus to Mars would be precipitous. The American continent would be divided, with a number of independent countries constantly squabbling among themselves for control of resources and territory. And if that happened, by circular argument, America would also be drawn into European power politics, caught up in the continent's imperial ambitions and internecine wars. In short, therefore, the United States came into being in part because of a fear that unless it did so it would both resemble contemporary Europe and be vulnerable to its influence.

It took Europe nearly 150 years to see the merit in Hamilton's argument. He had taken Europe as an example that America could avoid. From his American perspective, the contemporary European model of political organisation was hardly blessed. It demonstrated that nationalism – and imperial ambitions – led to war. The European Union, however clumsy in its construction and inefficient in its operation, offered the prospect of a supranational context in which some of the continent's nations – and in 1973 even the United Kingdom – might cooperate. The impetus towards greater European integration, originally presented as an economic impulse, but, for its advocates, always a political project, was thus also based on fear. This time the fear was that Europe would continue to be like Europe. After two world wars, many Europeans came to believe that enough was enough and that closer integration would be a way of breaking with the past.

The European project, like the United States, was the product of war. But there was an important difference. It was political glue that would first be used to bind the thirteen former colonies in America together, whereas the adhesive used in Western Europe was economic. One of the seminal events of the American War of Independence did not take place on the battlefield. Rather, on July 4 1776, the Continental Congress issued the Declaration of Independence, authored by Thomas Jefferson, and summarising the 'idea of America' as a democratic republic founded upon the principles of liberty and equality. What the founding period was about was the translation of these political ideas into a coherent framework of government: the ultimate achievement of the Philadelphia Convention in 1787.

The revolutionary impetus that led to the Constitution and the creation of the United States of America was not as cohesive a force as at first it might appear. Indeed the full import of Benjamin Franklin's famous remark, that the Convention had achieved 'a republic if you can keep it', became apparent soon after the Constitution was ratified. Federalists and Republicans argued over their competing interpretations of the meaning and nature of the revolution itself. For the Republicans, the 'idea of America' – a democratic federal republic – was sufficient as the political adhesive that would keep the United States united. On the other hand, the Federalists – perhaps more pragmatically – believed in the primacy of the Union as a strong centralised government that would be the economic engine for growth and prosperity. For them, it was economic as well as political glue that would make the Union work.

In practical terms, in the early years of the republic, such arguments would revolve around such key issues as the assumption by the federal government of state debts, and the competing visions on the constitutionality of a national bank expressed by Jefferson and Hamilton. When the Republican

Jefferson (soon the party's name would change to Democrat) defeated John Adams in the presidential election of 1800, the 'idea of America' as a primarily political construction appeared to have triumphed. Yet, as the subsequent sixty years demonstrated, the dispute between states' rights and federal authority, revolving around the issue of the retention and extension of slavery in the South – an economic argument as well as a political, cultural and moral one – proved to be an effective solvent for the 'bonds of Union'.

The Federalist ideal of a strong national government in which political and economic glue sustained the 'idea of America' thus ultimately was realised through the achievement of the first President who represented what subsequently endured as the Republican Party: Abraham Lincoln. The Civil War was fought to keep the United States united, and then the national ideal was consolidated through the motor of industrial capitalism. It was only after 1865 that both the political and the economic glue – symbolised by the establishment of a national currency – combined to strengthen a United States in which the issue of the boundaries of state and national sovereignty was more or less resolved (while admitting that in the South a stand-off continued until at least a century after Appomattox).

Can this American experience be mapped onto the contemporary situation across the Atlantic divide? After the Second World War, the fear of a continuing history of conflict gave an increased momentum to an 'idea of Europe'. It was a vision similar to that of the Federalists in America. Through closer economic integration, the political disputes that had previously characterised European history might be contained. From its origins, therefore, the European project, conceived as a political response to historical experience, relied upon economic glue as its principal adhesive.

The end of the Cold War had another impact upon this 'idea of Europe'. The re-unification of Germany altered the political framework within which the European Union could develop. The possibility of further integration – involving the admission of nations from the former Soviet Union – became part of the impetus towards the prospect of a United States of Europe. At the 'end of history', Europe's embrace of capitalism and democracy appeared to be at the forefront of modernity, despite its experience of conflict in the Balkans. The gathering pace of the drive for political union in an expanded European community became a post-Cold War phenomenon that was the product of the fading of that ideological conflict. Yet the outcome – the proposal for a European Constitution – became unstuck because, unlike the United States, thus far it had not found a common political glue strong enough to hold disparate states together.

In this respect, it is worth reflecting that after the controversy over the creation of a National Bank in America – it lost its charter during Andrew Jackson's presidency – there was little prospect of the United States adopting a common currency. Indeed, when the Civil War broke out, as Bray Hammond observes, the Lincoln administration had a problem. It was faced with 'the anomaly of a sovereign authority waging a war in defence of its sovereignty without possessing that most ancient and elementary attribute of sovereignty – control of the monetary system'.[19] In 1862 the federal government issued 'greenbacks', but even after the establishment of the dollar as the national currency, a central bank – the Federal Reserve – was established only in the early years of the twentieth century, since when it has assumed a place at the centre of fundamentalist conspiracy theories about the true nature of the American political system. The reluctance to establish national economic institutions – a bank and a currency – suggests the extent to which the 'idea of America' remained initially a political construction. This could only carry it so far. Despite the Founders' hopes, in 1860, war came. The United States thus took three-quarters of a century after its Constitutional Convention in Philadelphia to find that combination of political and economic glue that would serve to keep it united. This achievement was symbolised most eloquently in the Gettysburg Address, when Lincoln, the archetypal Federalist, harked back to the idealism of Jefferson's Declaration of Independence, even as his achievement in preserving the Union laid the foundations for the Gilded Age which followed.

As the history of the United States demonstrates, the problems faced by autonomous sovereignties that merge their distinctive identities in the hope of greater security will always emerge on the borders between state and federal authority. In America, slavery was the decisive issue, and the question of sovereignty was framed in terms of the states' rights to conduct their own political affairs, and ultimately to secede from the Union if they so chose. In Europe, the question of sovereignty has proven equally problematic, especially when it is seen, as it has been in the United Kingdom, in the context of the creation of a common European currency. With the contemporary failure to ratify the Constitution, the endurance of the Euro may nevertheless remain the talisman of the European project. The establishment of a common currency is testament to the faith that there is an economic glue that can bind different nations together. But the political glue is yet to be found. Just as Europe has to discover its Jefferson, its Madison and even its Washington if it is to achieve 'a union, if you can keep it', it will also need its *Federalist Papers* to persuade its peoples of the worth of that ambition.

Conclusion

From a European perspective, the political, economic and cultural development of America from the late eighteenth century onwards, from colonial status to independence, from agrarian subsistence to industrial capitalism, from frontier society to transcontinental urban sophistication, was a phenomenon that might inspire either shock or awe. As its experiment with republican democracy survived the trauma of Civil War, and the United

States began to emerge as a rival to European empires on the international stage, it was as difficult for Europeans to ignore America or to view the country with equanimity as it had been for the Abbé Raynal's contemporaries to remain neutral about the discovery of the new world.

Consider, then, how two of Europe's most influential twentieth-century thinkers might have responded to the Abbé's challenge. Biographically there is a clear contrast between them. For Albert Einstein, America was indeed a blessing. It became his home. In 1932, a month after Hitler came to power in Germany, Einstein left Berlin and settled in the United States in Princeton, becoming an American citizen in 1940. On the other hand, for Sigmund Freud, the United States was a different kind of asylum. Visiting in 1909, he proclaimed that America was, famously, 'a mistake; a gigantic mistake, it is true, but none the less a mistake'. From his European perspective,'America is the most grandiose experiment the world has seen, but, I am afraid, it is not going to be a success.' When Freud too was forced to leave continental Europe he lived out his life in Britain.

Freud rejected the idea of America whereas Einstein accepted it as a place of refuge. But as European intellectuals, like Condorcet before them, they both worried about the issue that nowadays has come to define different transatlantic perspectives on the wider world: the application of military force to political problems. In 1932, at the invitation of the League of Nations and through its International Institute of Intellectual Co-operation in Paris, Einstein initiated an exchange of letters with Freud, twenty-three years his senior, asking whether, in his view, there was 'any way of delivering mankind from the menace of war'. With Hitler taking power in Germany, Einstein reflected on the ability of political leaders to galvanise a population in support of war, and wondered if this was 'because man has within him a lust for hatred and destruction'. He argued that: 'in normal times this passion exists in a latent state, it emerges only in unusual circumstances; but it is a comparatively easy task to call it into play and raise it to the power of a collective psychosis'. He invited Freud to consider whether or not it was 'possible to control man's mental evolution so as to make him proof against the psychosis of hate and destructiveness' that had hitherto resulted in nations going to war.

Freud's reply accepted the fact 'that there is no likelihood of our being able to suppress humanity's aggressive tendencies' but he retained the hope that these could be sublimated through the 'cultural development of mankind' or the progress of civilisation. Since 'war runs most emphatically counter to the psychic disposition imposed on us by the growth of culture; we are therefore bound to resent war, to find it utterly intolerable'. Moreover, Freud hoped, 'with pacifists like us it is not merely an intellectual and affective repulsion, but a constitutional intolerance, an idiosyncrasy in its most drastic form. And it would seem that the aesthetic ignominies of warfare play almost as large a part in this repugnance as war's atrocities.'[20]

A year after this correspondence, Freud's books were burned in public by the Nazis. In 1937 he left Austria after its annexation. He died in London in September 1939, a matter of weeks after the Second World War had broken out in Europe. A month earlier, in America, Einstein had written to Franklin Roosevelt alerting him to the possibility of developing nuclear weapons. This kick-started the Manhattan project that in turn culminated in the dropping of atomic bombs on Hiroshima and Nagasaki.

In Europe, the trauma of the Second World War produced a popular reaction that would have met with Einstein's and Freud's approval. The continent made a purposeful effort to turn towards Venus. Eastern and Western Europe embarked on different roads to a similar objective: to avoid internal conflict between nations within the boundaries of their strategic influence. The Soviet Union held one part of the continent together through political and economic ties and, if necessary – in Hungary in 1956 and in Czechoslovakia in 1968 – military force. Western Europe, with a different ideological perspective and recognising an alternative tactical necessity, began the project, initially economically driven, but with a clear political agenda, to move toward closer integration. True, the Western European retreat from Empire was made sometimes and somewhat reluctantly and did involve military action overseas: the French war in Indo-China that culminated in defeat at Dien Bien Phu in 1954 is but one example; so too is the Anglo-French adventure at Suez in 1956. And during the Cold War the divided city of Berlin was a constant reminder that, in terms of a broader geopolitical perspective, the continent's ideological differences could still be the cause of potential conflict. Nevertheless, for the first time in its modern history, Europe avoided major wars.

At the same time, in the United States, the development of nuclear weapons as an outcome of Einstein's physics led to different political realities both internationally and domestically. Internationally, America became unquestionably the most powerful nation, and, domestically, the President, as commander-in-chief with the ultimate responsibility for deploying the nation's missiles, was, in Alfred de Grazia's apt phrase, 'the focus of the anxious crowd of the age' as the Cold War became the defining ideological conflict of the contemporary era.[21]

If presidential power was to be framed in terms of the Executive's responsibility to decide whether or not or when to unleash weapons of mass destruction, with all the apocalyptic consequences that might result, then it remained potential as long as rational political considerations held sway. The Cuban Missile Crisis, with its emphasis on brinkmanship and secret diplomacy, demonstrated John F. Kennedy's mature realisation that such power had to be tempered with responsibility and an awareness of the consequences of its use. Nevertheless, in the nuclear age, and during the Cold War, Americans initially accepted that the President as commander-in-chief could engage the nation in military activities that fell short of all-out atomic warfare. But presidential wars soon became politically controversial. President Truman in Korea, and more crucially Presidents Johnson and Nixon in Vietnam, fought wars which ultimately impacted upon American society, politics and culture – and the presidency itself – more profoundly than had any conflicts since the nation's own Civil War. The President's authority to commit military forces overseas was a potential usurpation of constitutional power, and not in keeping with the Founders' intent. Not only Americans but also Europeans could debate the appropriate limits of the Executive's authority as commander-in-chief.

In June 2000 Bill Clinton became the first American President ever to be given the Charlemagne Prize, the European honour awarded to those who are judged to have helped the cause of European unity and progress. Just over a year later, and eight months into his presidency, his successor was proving strikingly less popular in Europe. In August 2001, a Pew Research Center poll showed a remarkable consistency among those surveyed in France, Germany and Great Britain when they were asked whether they thought George W. Bush's understanding of Europe was less than that of his predecessors. Three out of four agreed that it was. Averaged across these three countries, 75 per cent of these people also thought that this President made decisions based only on US interests.[22]

If such sentiments were manifestations of broader anti-American feelings, which is in itself a contentious question, there is no doubt that in the immediate aftermath of 9/11, attitudes towards the United States, if not its President, changed. In France, *Le Monde* was moved to claim '*nous sommes tous Américains*'. The German Chancellor, Gerhard Schröder, offered 'unlimited solidarity' to the United States. In London the military band at Buckingham Palace played the American national anthem.[23] And in Washington, George W. Bush declared war on terror. For neo-conservatives in the administration the terrorist attacks both confirmed their world-view and presented an opportunity. America had been vulnerable because of its weak responses to threats from abroad, and its reluctance to use military power. Whatever the truth in the widespread rumours that members of the administration wanted to attack Iraq in the immediate aftermath of September 11 2001, the fact was that at that time the President had popular domestic support and international toleration – not least in Europe – for some form of military action.

For those who see presidential power as predicated on the American Executive's ability to use military force in the pursuit of national political ambitions, and for those who are still convinced that the 'idea of Europe' represents the way to break free from a history of rivalry and conflict, however, public opinion remains a formidable obstacle. From a consideration of the themes that underpin the contemporary debate on the relationship between Europe and America, indeed, it appears that the future divide for

Europe and for America is not so much the product of a transient ideological ocean between the continents, but rather the prospect of a permanent political disconnection between leaders' actions and popular reactions. On both sides of the Atlantic the fuel for this discontent is suspicion of centralised power, whether it is expressed in terms of scepticism about the nature of presidential wars or in relation to a constitution for Europe. American support for George W. Bush's 'war on terror', or at least its practical consequence – the continued national military presence in Iraq – has eroded steadily. So too has the confidence of many Europeans in the worth of its Presidents and Prime Ministers trying to write a constitution and construct a United States of Europe. The paradox is that European federalists may discover in America's history a political glue that can bind disparate sovereignties in a common cause: Jefferson's 'idea of America' as expressed in the Declaration of Independence. And those Americans for whom the Vietnam syndrome still resonates have reached similar conclusions as to the political efficacy of military interventions to those agreed by Europeans in the aftermath of the Second World War.

In 1992, in an echo of Freud as Henry Luce's 'American Century' drew to a close, the historian Arthur Schlesinger, Jr, could still pose the question 'Was America a mistake?' Commemorating the five-hundredth anniversary of Columbus's epic voyage that resulted in 'the most crucial of all encounters between Europe and the Americas', he also reflected on the Abbé's competition and predicted that 'the Academy of Lyons, if it chose to revive the Raynal prize, would be flooded with entries'.[24] If the post-September 11 period has been characterised by increasing turbulence in the relationships between some European nations and America and has resulted in political divisions within Europe and in the United States itself, it is not simply because the 'war on terror' has created fresh tensions. Rather, contemporary problems of presidential power and constitutional authority, war and sovereignty are more complex than neo-conservatives in America and neo-federalists in Europe might care to admit. Instead of debating the depth of the transatlantic divide, moreover, it may be that Europe and America can still learn from each other and their separate experiences in relation to those political, economic and cultural concerns that continue to present them with formidable challenges.

Notes

1 R. Kagan, *Paradise and Power* (London: Atlantic Books, 2003), p. 3.
2 D. Echeverria, 'Condorcet's "The Influence of the American Revolution on Europe"' (1786), *William and Mary Quarterly*, 25, 1, 1968, pp. 85–108, p. 100.
3 Kagan, *Paradise and Power*, p. 86.
4 See H.-J. Lusebrink and A. Mussard, *Avantages et désavantages de la découverte de l'Amérique* (Saint-Étienne: Université de Saint-Étienne, 1994).

5 'Debates in the Federal Convention of 1787 as reported by James Madison', in *Documents Illustrative of the Formation of the Union of the American States* (Washington DC: Government Printing Office, 1927), pp. 558–63.
6 Ibid., p. 621.
7 H.A. Washington, ed., *The Writings of Thomas Jefferson* (New York: H.W. Derby, 1861), letter sent from Paris, September 6 1789.
8 'Helvidius', no. 4, September 14 1793, in T. Mason, R. Rutland and J. Sisson, eds, *The Papers of James Madison*. Volume 15: *24 March 1793–20 April 1795* (Charlottesville: University Press of Virginia, 1985).
9 F. Fukuyama, 'The End of History', *The National Interest*, Summer 1989, pp. 3–18.
10 Kagan, *Paradise and Power*, p. 3.
11 George Bush, *Remarks to the American Legislative Exchange Council*, March 1 1991, and *Remarks on Assistance for Iraqi Refugees and a News Conference*, April 16 1991, George Bush Digital Library, http://bushlibrary.tamu.edu/
12 C. Krauthammer, 'Good Morning Vietnam', *Washington Post*, April 17 1991.
13 See 'Democrats Call for Rove to Apologize', *Washington Post*, June 24 2005.
14 Kagan, *Paradise and Power*, p. 11.
15 'Treaty Establishing a Constitution for Europe', *Official Journal of the European Union*, 47, December 16 2004.
16 Kagan, *Paradise and Power*, p. 3.
17 Quoted in G. Wood, *The Creation of the American Republic* (Chapel Hill: University of North Carolina Press, 1969), p. 467.
18 Alexander Hamilton, 'Federalist 6', in J. Madison, A. Hamilton and J. Jay, *The Federalist Papers* (Harmondsworth: Penguin, 1987).
19 B. Hammond, *Banks and Politics in America* (Princeton: Princeton University Press, 1957), p. 724.
20 O. Nathan and H. Norden, eds, *Einstein on Peace* (New York: Schocken Books, 1960), pp. 186–203.
21 A. de Grazia, 'The Myth of the President', in A. Wildavsky, ed., *The Presidency* (Boston: Little, Brown & Co., 1969), p. 65.
22 'Bush Unpopular in Europe, Seen as Unilateralist', Pew Research Center, August 15 2001.
23 Quoted in I. Daalder and J. Lindsay, *America Unbound* (Washington DC: Brookings Institution Press, 2003), p. 80.
24 A. Schlesinger, Jr, 'Was America a Mistake?', *Atlantic Monthly*, September 1992, pp. 16–30.

3

THE OLD AND THE NEW

Germany, Poland and the recalibration of transatlantic security relations

Kerry Longhurst

When the war in Iraq brought into focus acutely different perspectives amongst European states on US foreign policy, the intra-European division was manifest in a particularly stark fashion in Poland and Germany. Outside of France and the UK, nowhere were positions so opposite and so vociferously defended as in Warsaw and Berlin. German and Polish positions seemed to typify notions of 'old' and 'new' Europe and epitomise the idea of paradise and power. The Polish government's keen adherence to US policy, 'feisty' Atlanticism and support for the use of pre-emptive force, in particular, contrasted sharply with Chancellor Schröder's stalwart opposition and seeming disregard for Atlanticism, which had previously been a key tenet of (West) Germany's foreign policy since the inception of the Federal Republic. The radicalisation of US foreign policy was largely seen in Poland as an acceptable response and policy adaptation to new post-9/11 security challenges. In contrast, the tenor of Bush's policies was greeted by German elites and society at large first with scepticism and subsequently with outright condemnation, a reaction which has been sustained hitherto. As a result of this, on virtually every count Warsaw and Berlin held conflicting positions.

It will be argued in this chapter that Polish and German discord over US foreign policy and Iraq was a result of divergent strategic perspectives which were born of contrasting national histories and responses to the end of the Cold War, differences which came to light after the change in America's foreign policy after 9/11. Since then, while German–US relations have ebbed, the strategic relationship between Poland (as well as other East Central states) and Washington has become more vibrant. By focusing upon Polish and German policies towards Iraq and the particular domestic contexts and rationales behind policy choices, this chapter will consider the ways in which transatlantic relations are evolving and how relations between the United States and an enlarging Europe might further develop.

Germany and Poland: converging priorities in a reuniting Europe

Although the frameworks of transatlantic relations had been changing over time since the end of the Cold War, in retrospect the 1990s actually saw more continuity than change and a re-birth of the transatlantic partnership. The Clinton administration pursued an active European policy, led first and foremost by Madeleine Albright, and nurtured relations with both traditional and newfound allies in Europe. On the central objective of reuniting Europe, US foreign policy came together with that of Poland and Germany in particular, a meeting of minds which provided an important impulse for both NATO and European Union (EU) enlargements. Within this broader context of transatlantic unity on key European questions in the 1990s, Polish–German *rapprochement* was decisive. The historical antagonism which had previously characterised relations between German and Polish states since the eighteenth century was largely overcome after the end of the Cold War. German Unification within a European context, Polish–German reconciliation and the onset of the democratisation process in Poland provided a permissive environment within which cooperation could flourish. This new Polish–German 'hinge',[1] or 'community of interests' helped bring stability to Central Europe after 1989; Polish European policy was led by the notion of 'through Germany to Europe', and Germany's new Ostpolitik and backing for NATO and EU enlargements followed from its support for the consolidation of the Polish reform process and the stabilisation of Germany's eastern neighbourhood.[2] Such was the proximity of German and Polish security interests after 1989 that commentators began to speak of a new special relationship akin to the Franco-German alliance, also built on reconciliation, which had paved the way for West European integration after 1945. Poland as 'Germany's France to the East' would help bring about the more complete integration of the whole of Europe.

The shared Polish–German project for European reunification, and the widening of European institutions in particular, was underwritten by a mutual interest in maintaining US involvement in European affairs and upholding the indivisibility of transatlantic and European security. Both Poland and Germany cherished the crucial role played by American diplomacy in bringing about an end to Cold War divisions. The implosion of the Soviet bloc led to Poland's acquisition of sovereignty and German Unification, which were largely viewed in Warsaw and Berlin as successes of US foreign policy. Indeed, there appeared to be parallels in German and Polish Atlanticisms. As the original 'step-child' of American democracy, the prosperity and stability of the Federal Republic of Germany derived to a great extent from the US's role as both mentor and security guarantor during the Cold War. The United States and NATO became one of two pillars in West Germany's foreign policy (the other being France), a structure which enabled the Federal Republic to commit to a European vocation, as well as being compliant and in synch with US security thinking. In a not dissimilar way, as a young democracy the Third Polish Republic was also mentored by the United States. As a protégé of America or 'new model ally', Poland emerged, like West Germany after 1945, as an amenable and eager young apprentice.

The changing context of European security

The convergence of Polish and German priorities and Atlanticist credentials did not persist, however, beyond the immediate goals of securing Polish membership in NATO and the EU, objectives which had provided the 'motor' of the new German–Polish partnership. Already the intensity and details of EU accession negotiations, which had begun in 1998, placed a strain upon relations between Bonn/Berlin and Warsaw. This tension grew further in the aftermath of the 1998 federal elections in Germany, which brought a Red–Green coalition to power and signalled a re-thinking of aspects of German foreign policy, in the direction of 'normality' and assertiveness. Similarly, Polish foreign and security policy was evolving and gaining a newfound confidence and closeness with the United States as NATO membership moved onto the horizon. Moreover, the conviction held by all post-1989 Polish leaders, that only the United States and NATO could provide Poland with real security guarantees, was strengthened further through the experience of the Balkans, where Europeans proved to be largely ineffective. The implications of these developments in Germany and Poland for transatlantic relations were not particularly clear at the onset, but became increasingly more perceptible in the run up to September 11.

It was arguably the changing geopolitical context of European security in the late 1990s that prompted increasingly divergent responses from Germany and Poland. As noted above, NATO enlargement was a goal for both states and when Polish membership was achieved benefits were mutual. Through NATO enlargement Germany became surrounded by friends and allies and thus the historical dilemmas associated with being a frontier state were virtually dissolved. As a consequence, the importance of the US security guarantee and especially the nuclear umbrella lost a huge amount of its former significance. For Poland, however, as the new NATO frontier state, certain dilemmas remained. Although membership in the Alliance was unanimously regarded in Poland as the foreign policy success of the post-Cold War era, bringing to a close the acute insecurities of much of Poland's history, old uncertainties persisted in the shape of Russia, and a range of new insecurities emerged in Poland's eastern neighbourhood. Such differences meant that Warsaw and Berlin now held distinctly different conceptions of NATO's role. For Poland, the vitality of the Alliance and its importance for Polish security derived from the traditional military provisions of Article Five.[3] On the other hand, German thinking about NATO was becoming more honed in non-Article Five non-military functions and the Alliance's broader political role in the transformation and consolidation of reforms in Eastern Europe.[4]

The evolution of Polish and German security perspectives was also shaped by the further elaboration of the EU's foreign and security policy ambitions. Deeper integration in this area always was and remains a key German policy objective. The Europeanisation of foreign and security policy and the notion of pooling national sovereignty and resources in this area meet Germany's multilateral instincts and drive for more communitarisation of EU policies. From a German point of view an EU foreign and security policy would enable the EU to speak louder and more credibly on the world stage and would enhance the European pillar of NATO. While in the run up to EU accession Polish policy remained largely silent on the question of foreign and security policy, after the European Security and Defence Policy (ESDP) as the military arm of the Common Foreign and Security Policy (CFSP) was articulated in 1998, Warsaw's opposition was ignited. As an EU outsider, Polish diplomacy had to guard against being too critical and therefore damaging the accession process, while at the same time trying to shape policy developments in line with Polish preferences – which were at the opposite end of the spectrum to Germany's. Warsaw's vision of EU foreign and security policy stressed the importance of an inclusive approach (bringing the soon-to-be EU member states into the decision-making process, but excluding Russia), maintaining national sovereignty and intergovernmentalism as organising principles and finally ensuring that NATO's pre-eminence was not challenged. With such important factors in mind – namely contrasting geopolitical concerns and threat conceptions; different understandings of NATO's purpose, the role of the United States and the necessity of military force in European security; as well as opposing visions of what kind of international actor the EU should become – it was perhaps not unanticipated that after 9/11 Polish and German policies took such overtly diverging paths.

The radicalisation of US foreign policy: Polish and German differences exposed

After September 11, transatlantic relations and partners lost their centrality in US foreign policy; deterrence as an overall strategic doctrine was largely relinquished and was replaced with the right of pre-emptive military force. Unilateralism for the most part became an uncontested norm, a move which saw multilateralism disregarded and the Bush administration's disavowal of international regimes and agreements toughened. As a consequence, US policy towards Iraq proceeded on the basis of *ad hoc* agreements with individual 'willing' states, thereby undermining both NATO and the EU. The abrupt changes in US foreign policy in 2001 were greeted in very different ways in Germany and Poland. As already noted, increasingly divergent strategic perspectives, which had been emerging over time since the end of the Cold War, meant that in the face of 'naked US hegemony' Germany and Poland were predisposed to respond to America's call for a coalition of the willing in very different ways.

Poland – adhering to US leadership

It did not matter to the Polish government when the EU and NATO fell foul of US unilateralism in the war on terror; indeed, Poland emerged as one of the United States' key allies in the wake of 9/11. Polish diplomacy fell into line with US policy on virtually every count and in this context Poland's role as the US's protégé in the East was cemented. Polish support for US policy was fortified by the efforts of the UK, Germany and France to work out a coordinated agenda for the EU to respond to 9/11. Although this initiative came to little, the prospect of an exclusive Western European Directoire was unacceptable to the Polish government and helped affirm Warsaw's Atlanticism. The Polish fear was that, should the US lose interest in Europe, an exclusive club of privileged European states would seek to replace it and marginalise Poland's voice and influence.

Warsaw was quick to align with the US and to stress that Poland would not remain a passive participant in the war against terror. The events of September 11 also gave rise to a renewed impetus for Poland to act as a conduit for those countries in Eastern Europe aspiring to join Western institutions. In November Polish President Kwasniewski argued that Poland's place in the post-September 11 world order was to 'act as a leader to coax eastern nations into the Western camp and to persuade the West to accept them'. In this newly cast role as regional leader and friend of the US, Warsaw convened an Anti-Terrorism conference with leaders from Central, Eastern and Southeastern Europe, an event which demonstrated Poland's aspiring leadership qualities and commitment to the American-led campaign. Thus it was not surprising when there was little hesitation on the part of Kwasniewski to say yes to President Bush's request at the end of November for Polish troops to join the campaign in Afghanistan.

Having bagged the credentials of being one of America's principal allies in Afghanistan, when the US focus shifted to Iraq and regime change Warsaw's policy remained consonant with Washington's. Developments between the two countries since Afghanistan had had the effect of binding Polish foreign policy even closer to that of the United States. In late 2002 the US Congress approved a military loan of $3.8 billion dollars to Poland for the purchase of F-16 aircraft from American firm Lockheed Martin. Not only was this the largest single military loan in history, but because of the various lock-in effects the loan would cement the Polish defence sector to that of the United States for some time to come. Unsurprisingly, the Polish decision to 'go American' rather than purchase from the French or British–Swedish alternatives led to charges that Warsaw was not acting in a 'European spirit'.

Bitterness within Europe over the Polish procurement decision rumbled on into 2003 reaching a crescendo in the run up to war in Iraq. Although Polish participation in the Iraqi operation was by far the most sensational undertaking in Polish security policy of the last fifteen years, at the same time, the decision to participate was fully consistent with the pro-Americanism and Atlanticism of Polish post-1989 foreign policy. Polish Iraqi policy essentially flowed from its strategic relationship with the US and the belief at the heart of Polish thinking that the US remained the ultimate guarantor of Poland's security. Consequently, in the context of the emerging transatlantic and intra-European rift, Poland sided fully with the United States, and expressed this view alongside others in the 'letter of the eight' entitled 'Europe and America must stand united', published in January 2003.[5] Followed by the 'letter of the ten', these statements affirmed transatlantic solidarity between the signatories and the United States, helped sanction the US route to war in Iraq and unseated the Franco-German motor as the main force behind EU foreign policy. The letters also exposed wider fissures across Europe, further ignited by Donald Rumsfeld's reference to 'Old' Europe and 'New' Europe, a response to French President Jacques Chirac's lambasting of Poland and other East European states for supporting US policy. Ultimately what these spats demonstrated was that the perennial dispute within the European Union, between Atlanticists and Europeanists, lived on and had gained a new intensity and significance in the context of EU enlargement. Iraq also brought into focus the question of whether, as some quarters assumed, it should be the older and larger member states speaking for the EU, with one notion being that acceding states should accept a subservient backseat role in developing the EU's foreign affairs.[6]

Justifying going to war

Against the landscape of European disharmony the decision to send Polish troops to Iraq was made. In comparison to the United Kingdom and other contributing European states, the domestic context of debate and decision making over Iraq was rather muted in Poland. The key driving force behind Polish policy was to do with status and role and chiefly a desire to demonstrate Poland's loyalty as America's model ally. What characterised the Polish discussion on Iraq was its lack of detail and consideration of what Poland's interests in the region actually might be. Such issues were not debated at an elite level, and neither did they pervade a broader public discussion. Instead much was assumed and policy was led by reflexive Atlanticism and support for the United States. Significantly, there was no justification of Poland's involvement in the campaign in terms of responding to a direct threat. There simply was no suggestion that Iraq presented a 'clear and present' danger. Consequently, in contrast to the US and UK discourses, it was never envisaged that Polish territory was actually under threat by the hidden Iraqi nuclear arsenal. The Polish discourse was explicit about the need to preserve transatlantic bonds. Thus the notion ran that Poland's involvement in Iraq was crucial to prevent a serious split from emerging across the Atlantic as well as the US's withdrawal from Europe. In short, the Polish decision to deploy troops was explicitly linked to a desire on the part of Kwasniewski and the government to enhance Poland's Atlanticist profile as a serious global actor with international prestige.

The elite level consensus about going to Iraq was underpinned by a broader permissive environment. In the absence of a national discussion about Iraq, there was scant public interest and engagement on the issue. While public opinion was divided over the war elsewhere in Europe, Poland experienced no mass anti-war demonstrations and certainly nothing on a scale remotely comparable with other European supporters of Washington's policy, such as Britain or Spain. It was not therefore initially controversial when the decision was taken in June 2003 for Poland to stay on in Iraq to participate in the post-war stabilisation project. Moreover, the plan for Poland to have formal responsibility for one of the occupation zones was fully in keeping with the Polish decision to go to war in the first place and importantly took forward the goal of enhancing Poland's role and status as a serious security player and partner of the United States.

In the course of 2003 Polish policy became palpably less resolute and much more reflective. International criticism of Polish policy contributed to this and seemed to have been taken up a gear when one of the most ardent supporters of Poland's Atlanticist orientation, Zbigniew Brzezinski, expressed criticism regarding what he called 'a too-excessive and divisive demonstration of loyalty', which he regarded as damaging to Poland's relations with Germany and France.[7] The potency of this view, which added to a belief that Poland was fast becoming 'America's Trojan donkey',[8] took root in the context of the forthcoming enlargement of the EU.

It was perhaps when the first Polish casualty occurred in November 2003 that public opinion woke up to Iraq, and a broader consideration of the merits of the operation and Poland's role in it began to feature in party politics. These developments helped unleash a steady stream of calls for Poland to withdraw its troops. With the governing coalition weak and under stress, the question of the continuation of Poland's role in Iraq gained salience in the Sejm (Parliament) and the broader public arena. Such pressures ultimately compelled the government to pledge that it would reduce the size of the Polish contingent from the beginning of 2005 to around 1,500 troops, though at the same time it maintained that Poland would remain an occupying power until the full expiration of the UN mandate in December 2005. Over a year after Poland took responsibility for the occupation zone in Iraq much of the government's early confidence and optimism had subsided. Iraq continued to be highly unstable, a growing number of Polish troops were being killed and very few material benefits had materialised. In September 2004 the vast majority of Poles, over 70 per cent, wanted Polish troops to be fully pulled out of Iraq. By 2004 not only did public opinion arrive at the position of wanting the troops out, but it also came to support the notion that Europe's foreign policy profile should be boosted. In the context of a continuously bad situation in Iraq, negative public opinion was spurred on by two particular issues. First, the low level of involvement of Polish firms and businesses engaged in Iraqi reconstruction projects was a disappointment to a domestic audience keen to see some tangible benefits from participating in the US-led war. The second issue was that of visas. Despite Poland's close adherence to US policy since 2001, Poles became subject to the US government's decision to maintain and strengthen its visa regime, thus making it even harder for Polish people to travel to the United States. When Kwasniewski raised the visa issue with George W. Bush, it was made clear to him that his query was bordering on inappropriate and that no change of policy was going to transpire. Disappointed, Kwasniewski then argued that he was 'hurt' by the visa decision and that as 'a friend of America' he did not understand it. He also appealed for a more 'gracious' and 'less divisive' America.[9]

But perhaps even more disturbing were the ways in which developments in Iraq actually seemed to undermine the lines of justification which had led Polish policy in 2003. Polish loyalty to the US did not prevent a split in the transatlantic alliance nor did it sustain America's commitment to European security. Poland's plans to engage Germany and then NATO in its zone in northern Iraq in 2003 largely failed, leading to further tensions in transatlantic relations. In the meantime, the US announced plans for large-scale reductions in its military presence in Europe.

These points notwithstanding, it is important to note that international criticism and public disquiet over Poland's Iraqi policy did not bring about a significant change. While Spain did an about-turn on Iraq and was subsequently dubbed an 'occasional Atlanticist', the coordinates of Polish policy largely remained the same and strongly Atlanticist with a NATO-first policy. In addition, amongst its first foreign policy statements, the new governing coalition in Poland announced that depending on the outcome of talks with the US in December Polish troops might remain in Iraq, thus overruling the decision of the previous government. Of even greater significance, the new Polish foreign minister, Kazimierz Marcinkiewicz, acknowledged that serious negotiations with the American government had been under way about the terms of Poland's entry into America's missile defence system as the 'key European partner' of the United States.

September 11 – Germany's 'unconditional solidarity'

Germany's initial response to the terrorist attacks of September 11 2001 was 'unlimited solidarity' with the United States. Schröder's early declaration of solidarity was subsequently backed by firm cross-party support at home. Berlin's solidarity also extended to its full support for the US's invocation of NATO's Article Five. Germany's unlimited solidarity did not, however, translate into unconditional support for an immediate knee-jerk US military response. Crucially, at both elite and societal levels reticence towards the use of force and fear of US unilateralism pervaded the German debate. Consequently, Germany pursued its traditional preference for a multilateral approach explicitly aimed at tackling the roots of terrorism via political as well as military means. In this early phase Germany played an important role in the consolidation of an international alliance against terrorism. Foreign Minister Joschka Fischer in particular set about forging a common EU diplomatic response to the attacks on the United States and emboldening the role of the United Nations.

The US strategy towards Afghanistan, despite the invocation of Article Five, reflected a strong desire to lead and to forge a 'coalition of the willing' and it was in this context that Schröder's pledge of solidarity with the US was tested, when in November President Bush requested a military contribution. The Chancellor's response, which had already been articulated in a statement on October 11 pledging a military contribution to the war in Afghanistan, clashed with staunch domestic opposition. Schröder became caught in his own rhetoric. Aside from the PDS (the successor party to the SED) no party would openly condemn the US's right to pursue the perpetrators of September 11. At the same time there was no real appetite, not least from within the governing coalition itself, to engage German troops in an undesirable military campaign. Consequently, the Chancellor had to use all possible tactics if he was to fulfil his pledge to the United States.

On November 6 2001 Schröder announced that 3,900 Bundeswehr troops would be made available. To gain support for the contribution, proponents pointed to Germany's international responsibility, its role as a transatlantic partner, and the general credibility of German foreign policy, which was at stake. Proponents pointed too to UN resolution 1368, passed after September 11, condemning terrorism and recognising the right of nations to self-defence. Those opposed to the deployment pressed for restraint, more multilateralism and the inclusion of vital political and social measures into Operation Enduring Freedom. While the cabinet approved Schröder's plan for the Bundeswehr's deployment, which also met with broad approval from the opposition parties, support was far from forthcoming from substantial elements of the governing coalition. As a consequence of this the stability of the governing coalition was put under stress. The gravity of the issues at stake, combined with Schröder's weak domestic position, made it crucial to get the backing of his own coalition. The Chancellor subsequently decided to tie the issue of a Bundeswehr deployment to a vote of confidence in his government.

On Friday November 16 the Bundestag debated whether Germany should make available 3,900 Bundeswehr troops to participate in the war in Afghanistan and whether the SPD–Green coalition should remain in government. On these two issues Bundestag members were permitted one vote. The Vertrauensfrage was legitimate, Schröder argued, as the deployment issue was one of fundamental importance. Broad support was required, since Germany needed to show both at an international as well as a domestic level that the governing coalition was able and willing to back the deployment. On the second point, Schröder stressed the need for German foreign and security policy to be seen as consistent and in line with multilateralism, for Germany to be seen as a reliable ally, able and willing to make contributions to international security alongside allies and partners.

Extending the war on terror

The delicate domestic setting of Germany's contribution to Operation Enduring Freedom evolved further during the next half-year, taking German thinking even further away from that of the US (and Poland) about extending the scope of the war on terror. The German discourse remained focused on the need to address the underlying social and economic causes of terrorism and to do so without relying on a purely military approach. In contrast to the situation in Poland, early 2002 saw the beginning of the entrenchment of Germany's reluctance to extend its participation in America's anti-terror campaign. Afghanistan was one thing, but the notion of pre-emptive strikes on other states clashed head-on with German perspectives. The increasing belligerence of US policy, identification of 'rogue states' and articulation of the 'axis of evil' fanned the flames of mistrust which already existed in the Federal Republic.

The Munich Security Seminar in February 2002 became a platform for both Americans and Europeans (meaning primarily France and Germany) to air their views on the international situation. Unsurprisingly the war of words at Munich led to a firming-up of the contrasting European (again, primarily French and German) and US perspectives on how to proceed with the war on terror. Paul Wolfowitz berated the Europeans for their lack of military prowess and confirmed that the United States would, in the future, feel free to pick and choose its allies and partners, warning that NATO states should no longer consider themselves to be in a privileged position.[10] While the American delegates brought home the message to the Europeans, and in doing so clearly spelt out their agenda for Iraq, German defence minister Rudolf Scharping posited that there were no plans afoot for an invasion (of Iraq) and, furthermore, that it would be naïve to believe that Europe would support such military action.

Bush's identification of Iraq as part of the 'axis of evil' led the US administration to view it as the next battleground in its war on terror. Unsurprisingly this conclusion and the strategy of pre-emptive use of force that it implied collided with current thinking in a number of mostly West European capitals, but especially in Berlin. The German response to the US's expansion of the war against terror to Iraq was always going to be less than unenthusiastic. As already noted, the fragility of elite support for Afghanistan, combined with the overwhelmingly negative reception in Germany of US neo-conservative foreign policy, was a vital indicator for what became Schröder's subsequent inflexible Iraqi policy. Indeed, the clash with US foreign policy objectives could arguably be observed even in the early days after September 11 2001. Germany's 'unlimited solidarity' was coupled with a comment from Schröder that he would not let Germany participate in any 'adventures' and that, before coming to the aid of its allies, Germany would need to be fully consulted before the initiation of any military force.

Towards the end of summer 2002 the question of Iraq moved to centre stage. The strident rhetoric in the US which was pointing to a pre-emptive military strike in the near future had the effect of strengthening Schröder's resolve to give the situation and diplomatic means more time and, crucially, to allow the United Nations and not the US to decide how long the weapons inspectors should remain in Iraq. Denouncing the US's 'military adventurism', Schröder posited that a US-led war to oust Saddam Hussein would distract from the war against terrorism and would endanger the West's relations with the Islamic world. It was in this context that the notion of a Deutsches Weg or 'German way' was articulated in Germany as a means to describe a specific German approach to international affairs, but also to demonstrate to the United States that it would be Berlin's objectives and priorities that would determine the German stance on Iraq. Schröder's critics berated the idea of a 'German way', arguing that it sent out confusing messages to Germany's allies and, moreover, suggested the return of a dangerous Sonderweg, out of kilter with the Federal Republic's foreign policy tradition.[11] The transatlantic discussion, as it was, about extending the scope of the war on terror and the use of force towards Iraq was cut somewhat short when on August 26 2002 US Vice President Dick Cheney called for preventative military action to oust the regime of Saddam Hussein, an announcement which allegedly was not communicated to Germany in advance. Thus a wide-ranging debate was eclipsed, America's resolve to use force against Iraq, unless Saddam Hussein left the country, intensified, and Europe's divisions over the issue hardened, with Germany's position drifting even further from that of the US.

Weakened by a poor economic record and facing a strong challenge from the CDU/CSU led by Edmund Stoiber, Schröder seemed certain to lose the forthcoming federal election in September 2002. Seizing the initiative and responding to the widespread domestic reticence towards US policy and anti-war sentiment, the Chancellor affirmed his opposition to a war with Iraq, even if there were a UN mandate, and pledged to keep Germany out of any conflict. On September 22 the Red–Green coalition secured a victory, albeit the slimmest in the Federal Republic's history. However, Schröder's electoral success was not celebrated across the Atlantic. With reverberations still fresh in the air from the German justice minister's comparison of Bush's foreign policy endeavours with those of Adolf Hitler, US Secretary for Defence Donald Rumsfeld claimed that the Chancellor had irrevocably poisoned US–German relations.

The resolute stance taken by Schröder in September 2002 left little room for adaptation, modification or flexibility and crucially lost Germany any leverage that might still have existed to exact pressure upon Bush. Moreover, the effect was to isolate Berlin from wielding any real influence over other European partners. In this context it became apparent shortly after the election of September 2002 that Schröder's strategy had perhaps been a step too far. The German press were already decrying him for damaging American–German relations and, after a direct intervention from the US ambassador to Germany objecting to pervading anti-Americanism in the government, the Chancellor set about trying to pull Germany out of its self-inflicted isolation.

A temporary semblance of normality in relations was reached when Bush and Schröder declared their intention to get back to 'business as usual'. Berlin was keen to throw off the label of being seen as anti-American, but at the same time sought to reaffirm the view that still more time was needed to find a diplomatic solution to Iraq and that Germany would not participate in any military adventure. Despite this, by spring 2003 European perspectives towards Iraq had become rigidly polarised between 'hawks' and 'doves'. On January 20 Germany and France initiated a debate about terrorism, including the issue of Iraq, in the UN Security Council. The Franco-German initiative unleashed a wave of indignation across Europe. Discord revolved around the question of 'who speaks for Europe?'[12], while the Franco-German proposal at the UN claimed to be speaking in Europe's name, contrary voices proposed an opposing European discourse, which appeared in the form of the letters of the '8' and the '10'.

German policy remained grounded by the desire to avoid a military conflict. To this end, together with France and Russia, Germany pressed for an open-ended extension of the UN weapons inspectors' remit in Iraq and at the same time, in the context of NATO discussions, refused to support the formal authorisation of advance military planning, instigated by the US, for Turkey's defence. While Schröder was quick to assure Turkey that, if push came to shove, Germany would indeed come to Turkey's aid, the Chancellor was resolute in his opposition and along with France and Belgium argued that NATO should not pre-empt any decision by the UN Security Council on a second resolution. The rift over Turkey prompted commentators on both sides of the Atlantic to decry the death of multilateral security institutions. While the Bush administration placed the blame firmly on those European states, but particularly France and Germany, for their profound lack of support and for undermining NATO by failing to fulfil their alliance obligations, Germany saw that it was America's drive to invade Iraq, no matter what, and its consistent disregard for consultation which had undermined multilateralism.

From mid-February until the beginning of the invasion little ground was made up between the two camps. The Chancellor's pledge that he rejected German participation in any war, and his refusal to sign up to a UN resolution that permitted one, remained set in stone. Speaking in early April, Schröder defended the stance that he had so resolutely stuck to over the previous six months, confirming again that Germany would not take part in the war, and argued that there had been an alternative to war, 'but we were not able to prevent it'. The gap between US and German policies over Iraq was never truly reconciled, despite Schröder's claim that the US–German relationship was 'vital'. As a parting shot, a comment from the former US ambassador to Germany encapsulated the American government's view that Schröder's dispute with the US over Iraq 'was an opening shot of a new disloyal, confused and weak Germany'.[13]

A recalibration of transatlantic security relations?

A principal question posed by this book is whether a 'rift' has developed in the transatlantic relationship. The cases of Poland and Germany, examined through the prism of security policies in the period from 2001 to 2003, suggest that, rather than a 'rift', a recalibration or adjustment in transatlantic relations has and is still taking place. Poland and Germany are both key actors in this process of change and provide perhaps the best illustrations of the varied forms of Atlanticism which have emerged over the course of time within Europe. Poland's take on the transatlantic partnership and the role the United States should play in European security and that of Germany represent contrasting readings of and responses to the end of the Cold War and of course the profound changes in US foreign policy after 9/11. These divergent responses, echoed elsewhere in Europe too, will provide the contours and substance of a more fluid and less internally cohesive transatlantic partnership with diverging strategic priorities.

Much of this depends, of course, on the future direction of American foreign policy. Although in its second term the Bush administration has sought to lay a greater emphasis upon multilateralism and America's

European partners, unilateralism and the right of pre-emptive force will remain key tenets of US policy. A further possible change in US foreign policy, with sizable implications for transatlantic relations, is a drift into a more isolationist stance. Ongoing difficulties in Iraq, together with the effects of hurricane Katrina, have shaken domestic confidence in US foreign policy, the government, and the President in particular, as a man who 'gets things done'. A subsequent combination of US isolationism coupled with sustained unilateralism and belief in the right of pre-emptive strikes would not be good news for Poland, Germany or the wider Euroatlantic community. Beyond this longer-term prognosis, there are a number of observations to make about transatlantic relations and how and why they are being recalibrated in the light of Polish and German perspectives and policies as cases in point.

The dual enlargements of the EU and NATO: reinvigorating Atlanticism

The enlargements of the EU (2004) and NATO (1999, 2004) brought into the Euroatlantic community not just more member states and partners but more diversity. East Central European states brought with them differing strategic priorities, conceptions of sovereignty, attitudes to the use of military force and perceptions of the US's role in European affairs. These deep-seated preferences clashed with the strategic cultures and perspectives of a number of existing member states, but at the same time gelled with the interests of others. Poland is a prime example of a new NATO and EU member which brought in new perspectives and made an impact, but there are others too. The foreign policies of the Baltic states are even more Atlanticist than Poland's, and Romania and Bulgaria as soon-to-be EU members have also developed intimate relations with the US. Thus, EU and NATO enlargements have shifted the centre of gravity of the transatlantic relationship eastwards and in doing so have reinvigorated Atlanticism. This is to be seen most vividly in the recent realignment of US military policy in Europe; some troops previously stationed in Germany are to be relocated to Bulgaria and Romania, a shift which emphasises the apparent change in US strategic thinking, towards the Black Sea region and beyond.

It is perhaps remarkable that, despite the divisions in Europe over Iraq, the EU has made notable progress in the areas of CFSP and ESDP. The EU as a security actor has performed 'on the ground' in a variety of military and police missions in Europe and beyond, while through a harmonisation of French, German and British perspectives the EU has proved to be effective *vis-à-vis* Iran. There are, though, limits to what the EU can and wants to do as a collective actor. Within an EU of twenty-five members a lack of consensus prevails as to what type of security actor the EU should ultimately evolve into. The recent enlargement of 2004, coupled with the failed French (and Dutch) referenda on the EU constitution, substantially strengthened the caucus of states that (like Poland and the UK) at least on military issues seek to maintain a NATO-first policy rather than a 'less-America, more Europe' stance.

A more nuanced picture

The case of Germany demonstrates how in the absence of a Cold War threat the need for strict adherence to US policy became both questioned and jaded. This change was bolstered by generational change in Germany which brought a new generation of leaders to power which was less tied to German history and thus more at will to pursue a more confident and assertive foreign policy. Meanwhile, the example of Poland illustrates the vibrancy of the US's role and transatlantic relations in relation to a state only very recently emancipated from the confines of the Cold War. The passion contained in Polish Atlanticism, though still undoubtedly very strong, will change over time and will become more 'Europeanised' now that Poland is a full member of the EU. The positive effect of EU membership upon Polish policy towards the Ukrainian revolution in late 2004 brought into focus for Warsaw the way in which having EU backing visibly strengthened Poland's voice and objectives in international affairs.

A deduction which is often made is that Europe is and will remain divided on key strategic issues. It is logical that Iraq is posited as showing the degree to which Europe is still split between 'European' and 'Atlanticist' viewpoints. However, it is argued here that the traditional depiction of Europeanist versus Atlanticist credentials is not a nuanced enough framework to capture the ways in which the transatlantic relationship is currently evolving, and the complex reasons behind this.

Notes

1 Adrian Hyde-Price (2000) 'Stable Peace in Mitteleuropa: The German–Polish Hinge', in Arie Kacowicz, Yaacov Bar-Siman-Tov, Ole Elgstrom and Magnus Jerneck (eds) *Stable Peace Among Nations* (Boulder, Colo.: Rowman and Littlefield), pp. 257–76.
2 See Volker Rühe (1993) 'Shaping Euro-Atlantic Policies: Grand Strategy for a New Era', *Survival* 35, 2, pp. 29–137. Jaroslaw Drozd (2001) 'Polish–German Relations and the Question of Security', in Roman Kuzniar (ed.) *Poland's Security Policy 1989–2000* (Warsaw: Scholar Publishing House), p. 111; Adrian Hyde-Price (2000) *Germany and European Order: Enlarging NATO and the EU* (Manchester: Manchester University Press), pp. 149–54.
3 Kai Olaf Lang (2003) 'The German–Polish Security Partnership within the Transatlantic Context: Divergence or Convergence?' in Marcin Zaborowski and David H. Dunn (eds) *Poland – A New Power in Transatlantic Security* (London: Frank Cass), pp. 108–9.
4 Hyde-Price, *Germany and European Order*, p. 143.

5 *The Wall Street Journal*, January 30 2003.
6 Simon Duke (2004) 'The Enlarged EU and the CFSP', *Reports and Analyses*, April 5, Centre for International Relations, Warsaw, p. 4.
7 Interview with Zbigniew Brzezinski, 'Przez glupote i fanatyzm', *Gazeta Wyborcza*, October 7 2003; see also 'Zbigniew Brzezinski dla Rzeczpospolitej', *Rzeczpospolita*, April 20 2003.
8 'Is Poland America's Donkey or Could it Become NATO's Horse?' *The Economist*, May 10 2003.
9 'Polish President Appeals for a More "Open and Gracious" US', *The New York Times*, September 4 2004.
10 Elizabeth Pond (2004) *Friendly Fire: The Near-Death of the Transatlantic Alliance* (Washington, DC: EUS-Brookings Institute Press), pp. 1–3.
11 See Peter Rudolf (2005) 'The Myth of the "German Way": German Foreign Policy and Transatlantic Relations', *Survival*, 47, 1, pp. 133–52.
12 See 'Who Speaks for Europe', *The Economist*, February 8 2003.
13 'We're Not Children!' *The Economist*, May 17 2003, p. 38.

4

FROM DISPUTE TO CONSENSUS

The emergence of transatlantic opposition to neo-conservatism

Jean-marie Ruiz

There was a time, not so long ago, when American ideas about international relations, even those that were rather unflattering for the Old Continent, raised little opposition in Europe. The world seemed to accept that as the winner of a bipolar conflict the United States had a natural right to monopolize the analysis and interpretation of the post-Cold War world and its dangers. Francis Fukuyama, with his 'end of history', and Samuel Huntington, with his 'clash of civilizations', provided two antithetical visions of what the world would be like after the victory of capitalism and the collapse of communist ideology. In the absence of any real alternatives, these two paradigms provided the starting point for discussions and debates on both sides of the Atlantic. The fact that both these articles were written during a period of relative political consensus between the United States and Europe, especially compared to the period that was to follow, was undoubtedly an important factor in the favourable reception they were given. When political relations are set fair, American theorists dealing with international relations or Euro-American relations do not come up against major opposition in Europe. Even the ideas espoused by Kagan in 'Power and Weakness' were at first received positively, though they were often highly condescending towards the Old Continent.[1] Europe seemed happy to be regarded as 'weak' by a representative of the major 'power'. Kagan's views were certainly attractive to Europeans, who, in general, recognized themselves in the broad-brush portrait painted by the author. Accepting a certain decline, wanting to enjoy the benefits of modern society to the full without worrying about the order or disorder of the world, and expressing a preference for materialist and bourgeois values over martial values and the primacy of internal politics over foreign policy were not regarded as dishonourable by most European readers. After all, if the United States wanted to monopolize power, the Europeans, who were aware of the costs of wielding power, were very happy for them to do so.

When Europeans realized the sort of policies the neo-conservative ideas of Kagan and others would lead to, their reaction became more virulent. The deterioration in political relations between Europe and the United States was thus concomitant with an increase in intellectual resistance to these ideas. Unlike Fukuyama's and Huntington's writings, which had little influence on United States foreign policy, it is common knowledge that, in the wake of the September 11 terrorist attacks, President Bush's first term in office was guided by a neo-conservative agenda. It is also certain that transatlantic relations have been damaged by the ascendance of neo-conservative ideas within American circles of power, with the greatest friction being caused by the neo-conservatives' nationalistic outlook, their desire to perpetuate the United States' dominant position and their resolve to free America from the constraints of international treaties and organizations. The messianism and nationalism that could be seen in the Reagan administration's approach to diplomacy have been revived by the neo-conservatives, for whom they have been raised to the rank of doctrine. This approach is just as incompatible with European diplomacy as it is with European public opinion, which is automatically hostile to any 'neo-imperialistic' policy, even if it is 'benevolent', as Robert Kagan asserted in 1998 in an article in *Foreign Policy*. More generally, United States activism on a unipolar and American-dominated international scene is unlikely to be easily accepted by Europe. This is all the more the case in a context that is also characterized by an ever-widening transatlantic divide in terms of values and the evolution of society and mentalities. In other words, the values and principles of the neo-conservatives are too specifically American to strike a chord on the other side of the Atlantic. The new foreign policy doctrine that was drawn up following the September 11 terrorist attacks is stamped with the hallmark of the neo-conservatives. Largely based on the principle of pre-emption and with the explicit objective of maintaining the hegemony of the United States, and thus of a unipolar world, this doctrine is a serious challenge to the political principles to which the European states are deeply attached. Unlike the policy of containment that held sway until the end of the Cold War, the doctrine of pre-emption does not respect the sovereignty of states, despite this sovereignty being recognized and defended by the United Nations. The current American administration gives the impression that it will only respect the sovereignty of those states that are capable of taking action against terrorism; however, American neo-conservatives are quick to reaffirm their own sovereignty (most notably by refusing to be bound by international agreements, such as the ABM treaty and the International Court of Justice) and their right to act unilaterally.

Thus, it has been clearly established that the deterioration in transatlantic relations was a direct result of the rise of the neo-conservatives and the new approach to international relations and foreign policy that was adopted by the American government in the wake of the September 11 attacks. The strong opposition of many European countries to the subsequent war in Iraq was a reflection of transatlantic divergences in the way in which the contemporary world is perceived, in the best way to deal with new threats and in defining the United States' role in the world. France's position at the head of the anti-war movement reflects diametrically opposed points of view on how the post-Cold War world should be structured and on how international order and peace should be achieved. Consequently, France has become the *bête noire* of the neo-conservative movement.

Criticism of the ideas of the neo-conservatives, whose power seems to be declining at the moment is now coming from both sides of the Atlantic. This criticism may be examined by comparing Pierre Hassner's and Jean-Marc Ferry's critiques of Kagan's ideas (his article 'Power and Weakness' is both emblematic of the neo-conservative point of view and centred around European–American relations) with the arguments currently being used by opponents of neo-conservatism and moderate neo-conservatives in their debate with hardliners. The striking similarities between the arguments being put forward by critics in the United States and those in Europe suggest that Kagan's assertion that there is a profound difference between American and European perspectives is untenable. On the contrary, this common ground demonstrates the existence of a transatlantic community of thought.

Two early French criticisms of Kagan's ideas

Some of the staunchest criticism of Kagan's ideas was produced by proponents of European unification, who felt that Kagan had no real understanding of the nature and objectives of the European Union. As experts in the history of the United States and Europe, and in political philosophy, these critics were able to see through Kagan's sweeping generalizations and misinterpretations of other philosophers' work. Kagan's earliest critics included the French philosophers Jean-Marc Ferry and Pierre Hassner, two internationally respected writers whose arguments have since been taken up by others across the Atlantic, including Francis Fukuyama, who claims to belong to the neo-conservative movement. As the author of an important treatise on the European Union,[2] Jean-Marc Ferry immediately recognized the flaws in Kagan's work, both in terms of his misrepresentation of philosophical ideas and his poor understanding of European institutions. Similarly, Pierre Hassner, an acknowledged expert on American and European political thought and on transatlantic relations, had no trouble demonstrating the over-simplification and conceptual errors inherent in Kagan's writings. Although Ferry and Hassner are both proponents of European political union, they approached Kagan's work from different perspectives: Ferry focused on the differences between contemporary Europe and the United

States, whereas Hassner focused on the common ground between the two societies. The resulting analyses and the insight they give into 'Power and Weakness' are thus particularly interesting.

Ferry expressed his views in an interview in May 2003.[3] He challenged Kagan's central premise that transatlantic disagreements over international relations and foreign policy were essentially due to the relative strengths of the United States and Europe: that is, as Europe is weak, it favours actions that are coherent with this weakness. 'The explanation based on weakness is weak', writes Ferry, 'it does not allow us to understand how Europe manages to implement, by political means, a particularly advanced concept of the world order and of the setting up of laws that go beyond classic international law.' Far from being conditioned by factors that are partly independent of the will of current governments, these disagreements reflect deep-seated differences based on a deliberate political direction that leads to a different way of considering international relations. Unlike the United States, Europe has a favourable attitude towards what Kant called 'cosmopolitan law', which is fundamentally distinct from classic international law in that it is not based on the sovereignty of states but on the 'rights of people'. In this respect, Europe is in greater agreement with the evolution of international thinking and is, in some ways, at the forefront of a new direction for international law. A good example of this is the International Criminal Court, which penalizes human rights abuses at a trans-national level, rather than at a national level. By maintaining the primacy of sovereignty – at least of their own (for example, by refusing to ratify the International Criminal Court Charter) – and by opposing reforms of the United Nations that would lead to greater recognition of the rights of individuals and peoples at the expense of the sacrosanct sovereignty of states, the United States is turning its back on this evolution. More generally, the concepts of law and legality are viewed differently on the two sides of the Atlantic in that the United States has a tendency to confuse legality and morality, whereas, according to Ferry, Europeans are highly unlikely to confound the two. For example, the Bush administration invoked human rights and moral values to legitimize the war in Iraq, even though the war was illegal under international law (as was pointed out by the General Secretary of the United Nations himself):

> If Robert Kagan had written a book that addressed the substance of the matter, rather than focusing on Europe's military weakness, he would have highlighted the decisive question of the relationship between law, morality and religion, by comparing the American vision of the world (as it is presented by George W. Bush) and that of 'Europe' (as it is presented by France, Germany and Belgium in particular).

> This, it appears, is a key to the divergence between the two points of view. If human rights are defined in terms of morality instead of law, it becomes possible to invoke these rights to legitimize a 'just' war, even if it is unlawful. On the contrary, if, as should be the case, human rights are considered to be rights similar to all others and therefore subject to endorsement – preferably by the courts or through other procedures (in this case those of the United Nations Security Council and General Assembly) – it is no longer possible to cite human rights as justification for an action that is illegal.[4]

On the other hand, the European Union has a fundamentally different vision of power from the United States. This vision is not, as claimed by Kagan, dictated by Europe's weakness, but rather by its history and the political philosophy that has formed as a result of this history. Europe does not associate power with recourse to violence, and does not measure power merely in terms of physical strength: a state's influence on the international scene is not regarded as being proportional to its military strength. This idea is compatible with the European tradition of trying to maintain the balance of power and of using diplomacy (whose objective is to avoid war and the use of force) to resolve disputes. It is also coherent with the lessons of imperialism, which have clearly established that a dominant power, however strong it may be, cannot survive for long if it is not founded upon a sense of justice and if it ignores the love of freedom of the people who are subject to its grip. Ferry contends that Europe's reservations about the American neo-conservatives' avowed objective of democratizing the Middle East by creating the institutions needed to guarantee the rule of law are the result of its history and the lessons of imperialism:

> Americans are 'substantialist' in that they believe the way a democracy is established is unimportant. For them, all that counts is the creation of representative institutions [. . .] They want to create a system in which all the different ethnic groups and religions are represented and to set up a pluralist system that more or less resembles a state governed by laws and that is, of course, subject to free elections. Their philosophical error is to assume that reason may be imposed by force, even if the (popular) will is lacking. [. . .]
>
> Today Americans want to do what the French wanted to do two centuries ago. They want to export their democracy without bothering with the intricate relations between life, procedures, backgrounds, sensibilities and pride of peoples. Although the majority of Iraqis were hostile to Saddam Hussein and his regime, and even though, like people everywhere, they wanted to be free, it has become clear that this freedom can neither be imposed by force nor created overnight.[5]

Despite the painful experience of Vietnam, the United States' traditional lack of sensitivity to history, their refusal to take into account the role of peoples and their lack of self-criticism are all obstacles in the path of their claim to be a model. On the other hand, according to Ferry, Europe is at the forefront of a new, cosmopolitan and multi-polar international order. This is not only due to its respect for individuals and peoples, its more modern concept of power and sovereignty and its greater sensitivity to the lessons of history, but also to its greater respect for states, which it considers as indispensable for democracy, as well as to its belief that trans-national economic forces must be controlled and remain subordinate to political objectives.

Ferry's criticism of Kagan's ideas is designed to show that far from being more or less consciously guided by its weakness – of which its foreign policy is only a form of rationalization – Europe is actively trying to modernize international relations and international law along the lines of Kantian 'cosmopolitanism'. Contrary to Kagan's claim, this is not at all outmoded, or 'post-modern'. As it is more in line with the evolution of the world and the aspiration of its peoples, Europe provides a better model than the United States. The so-called 'weakness' of Europe is its strength, just as the so-called 'strength' of the United States is in reality an obstacle to its prestige on the international stage and to its sustainable domination of world affairs.

Similar ideas can be found in Pierre Hassner's criticisms, in which he denounces Kagan's misreading of earlier philosophical works, contests his analysis of American and European societies and values, and repudiates his definition of power.[6] Kagan's comparison of a 'Kantian' Europe with a 'Hobbesian' United States shows a deeply flawed analysis of these two philosophers' works. Far from forming an apology for the state of war and pagan virtues, Hobbes argues that the quickest way to end such strife is to submit individuals to a common authority. As Leo Strauss has established, Hobbes's ideas are more compatible with the construction of a world State than with the maintenance of an 'anarchic' international system. On the contrary, Kant – who according to Kagan was in favour of the immediate creation of a world State – never thought that everlasting peace would come tomorrow. In fact, far from affirming that peace would come from the immediate submission of states and peoples to the law and to reason, Kant believed that human progress and the road to peace would be punctuated by wars and disputes. Therefore, Kagan's vision should have led him to conclude that the United States is the 'Kantian' society and that Europe is 'Hobbesian'. However, Kagan's entire premise should be treated with caution as it is based on a very tendentious and selective analysis of the history of the United States and Europe. Contrary to what is affirmed in 'Power and Weakness', Europe has not always been in favour of 'appeasing' aggressive powers, and legalism and the cult of 'zero fatalities' are more characteristic of the United States than of Europe.

What Hassner wanted to show above all was that there is not one single Europe, nor one single America. Both entities are multi-faceted and constantly subject to internal divergences. On each side of the Atlantic there are varied points of view which, depending on the period and on events, influence political decisions. 'There are other divisions that often transcend the divergences between Europe and America', he writes. 'At certain times, majority opinions crystallize; at other times, they fall apart.'[7] In the history of the United States, it is easy to find policies that Kagan defined as being specifically European. For example, Wilson and Carter (whose positions cannot be explained by the 'weakness' of the United States) were in favour of a foreign policy based on international law. In reality, the ideas that are currently dominant in the circles of power in Washington are the result of the events of September 11 2001, and statistics show that they are far from being universally popular with the American people: a majority of Americans are not at all hostile to the United Nations and prefer to act in conjunction with their allies. The neo-conservatives also have their opponents on the political chessboard and in American intellectual circles; there is no need to cross the Atlantic to find diverging points of view about international relations and United States foreign policy. Many American opponents of the policy currently being carried out share the views of European critics. For example, Joseph Nye, the originator of 'soft power' (the ability to influence other nations by the power of persuasion and seduction), regrets the Bush administration's penchant for basing its foreign policy uniquely on military or economic 'hard power', that is, force. 'Theoretical discussions about power are not only as old as the theory of international relations (they are at the centre of the earliest classic work on international relations – Hans Morgenthau's *Politics Among Nations*)', Hassner writes, 'they are as old as the history and philosophy of politics itself'.[8] Thus, it is even more surprising that Kagan equates power with military power throughout *Power and Weakness* and that he never entertains the slightest doubt about the effectiveness of this military power. Similar views are held by the proponents of a more multilateral approach and, more generally, by all those, from Kissinger to Clinton, who insist on the limits of the United States' power. Furthermore, Hassner points out that the debate between proponents of the use of this power to spread American values (the idealists) and those who are more sceptical of the chances of success and the justification of such an approach (the realists) has existed throughout the history of the United States. Of course, this debate also exists in Europe, as Europe's leaders and influential commentators are not all supporters of pacifism and multilateralism. For example, Dominique de Villepin, who is often vilified by the neo-conservatives, believes as much in power as in international law. Care should be taken not to over-simplify matters and it should be accepted that the respective positions of Europe and the United States are more dependent on the political context, interests and dominant theories than on their respective power.

That being said, are there deeper and more permanent divergences between the ways in which international relations are seen on either side of the Atlantic? Have the wide differences in history and geography made their mark on this vision? As an expert on the respective histories and mentalities of each side, Hassner admits that certain generalizations can be made and that these generalizations can be used to draw up a structural contrast between Europe and the United States. Although this is particularly true for the neo-conservatives and for supporters of President Bush, most Americans believe that the United States has a special 'mission' in the world, because this notion has been firmly anchored in the collective subconscious since the creation of the Federal Republic. This 'exceptionalism' has given a universal character to their values and a legitimacy to the United States' actions around the world. As the United States Constitution is seen as a depository of universal values, actions that are seen to be legitimate from a national standpoint are often considered, a priori, to be legitimate from an international point of view. This aspect of the national psyche, which has been reinforced by the religious and nationalist revival of the 1990s and by the September 11 terrorist attacks, Hassner argues, goes against the dominant psychology in Europe and forms a durable basis for transatlantic differences, especially when American leaders use it to legitimize a unilateral foreign policy:

> It is that, as much as their belief in their material superiority, that explains America's amazing capacity to apply double standards to themselves and to the rest of the world. The United States can be altogether sovereignist when considering their own position and completely interventionist towards others. For Americans it is unthinkable for one of their own citizens to be tried before an international tribunal, but it is perfectly acceptable for America to put a price on the heads of foreign leaders, kidnap them and try them, or to execute Mexicans without allowing them access to their consular authorities, or to hold the Guantanamo prisoners in a judicial vacuum. It is because the United States believes both in 'American exceptionalism' and in the United States' universal value that it is so prone to misjudging its own nationalism and that of others.[9]

However, in contrast to Ferry, Hassner tends to stress what the two sides of the Atlantic have in common, rather than what separates them. He does not share the often-expressed idea of an opposition between a multilateralist Europe and a unilateralist America. On the contrary, he insists on the need to reconcile the two approaches, as a state's foreign policy cannot be based on either of the two approaches in isolation. Similarly, all states have to recognize that the world is multi-polar in some domains and unipolar in others. Above all, it would be best if each side could see that their differences are relatively insignificant compared with the common danger that threatens not only these two parts of the world, but the future of all humanity: the increasing strength of anarchy (which Hassner has long warned against by speaking of a return to the Middle Ages). In this respect, what Europeans should fear the most is not unipolarity, but 'apolarity', that is, the inability of the major powers to face up to new dangers, such as terrorism:

> The real challenge facing both the United States and Europe is the need to address the powerlessness of world powers to respond to world disorder, and the threats posed by networks of fanatics, by the mafias, by weak states and by uncontrollable societies. What we have to be wary of, much more than unilateral policies, multilateral organizations, unipolar or multipolar systems, are what Niall Ferguson has called 'apolarity' or 'generalized impotence'. Survivors of the 21st century, if there are any, may look upon the United States' imperial designs, France's national nostalgia and the pacific illusions of parts of Europe, with some irony, considering them all equally anachronistic in the face of the rise of anarchy.[10]

The American echo

American criticism of neo-conservatism is not, of course, new. From the very beginning, some criticisms were similar to those put forward by Hassner and Ferry. However, what is remarkable and significant about the revival of nationalism in the United States is that the self-righteousness underlying neo-conservative theories, which would have been more fiercely criticized at other times, passed virtually unnoticed in the particular atmosphere that reigned after the September 11 terrorist attacks. George Kennan, who often described himself as a foreigner in his own culture, was, in this respect, the exception that proves the rule. Before his death in March 2005, he actively denounced the pretensions of his country after the Cold War and he frequently mocked the Republicans' eulogistic speeches on their so-called victory over communism.[11] The last interviews he gave, in 1999 and 2002, show his total disgust with the new strategic doctrine, with the principles of the neo-conservatives, and with the United States' new diplomatic style, which Walter Mead dubbed 'Jacksonian' and which Michael Lind called 'the Texan spirit'.[12] Kennan also expressed his opposition to the Iraq war, using similar arguments to those used in France:

> The apparently imminent use of American armed forces to drive Saddam Hussein from power, from what I know of our government's

> state of preparedness for such an involvement, seems to me well out of proportion to the dangers involved. I have seen no evidence that we have any realistic plans for dealing with the great state of confusion in Iraqian affairs which would presumably follow even after the successful elimination of the dictator. . . . I, of course, am not well informed. But I fear that any attempt on our part to confront that latent situation by military means alone could easily serve to aggravate it rather than alleviate it.[13]

For such criticisms to be widely considered and cited in the debate on whether the United States' foreign policy is well-founded, they must be borne out by events and confirmed by certain American specialists, including Francis Fukuyama, the famous author of the 'end of history' who has always identified himself with neo-conservative circles. The invasion of Iraq in March 2003 not only led to a transatlantic dispute, it also created durable divisions within the neo-conservative movement. The magazine *The National Interest*, which has often served as a mouthpiece for neo-conservatism, was at the centre of a polemic between moderate neo-conservatives, who favour a foreign policy oriented towards the promotion of democracy around the world but who are not bound to an imperialist or unilateralist outlook, and the more messianic and imperialist neo-conservatives, who are prepared to use the United States' military might to democratize without worrying about international legitimacy.[14] This debate notably opposed Fukuyama, representing the moderate wing, and Charles Krauthammer, one of the most doctrinaire *neocons* and proponents of 'American uni-polarity'.[15] In an article entitled 'The Neoconservative Moment', published in 2004, Fukuyama distanced himself from what Krauthammer calls 'Democratic Realism', which combines all of the excesses of the extreme wing of the neo-conservative movement:

> [Krauthammer's] 2004 speech is strangely disconnected from reality. Reading Krauthammer, one gets the impression that the Iraq War – the archetypical application of American unipolarity – had been an unqualified success, with all of the assumptions and expectations on which the war had been based fully vindicated. There is not the slightest nod towards the new empirical facts that have emerged in the last year or so: the failure to find weapons of mass destruction in Iraq, the virulent and steadily mounting anti-Americanism throughout the Middle East, the growing insurgency in Iraq, the fact that no strong democratic leadership had emerged there, the enormous financial and growing human cost of the war, the failure to leverage the war to make progress on the Israeli–Palestinian front, and the fact that America's fellow democratic allies had by and large failed to fall in line and legitimate actions ex post.[16]

The fact that Fukuyama's arguments are very similar to those evoked above shows that the internal divergences within the United States are just as large as those that divide America from Europe and perfectly illustrate Hassner's premise.

There are several levels to these criticisms. Like Ferry, Fukuyama insists on the need to take into account the lessons of history and notes that in the past the United States has often failed in what they call 'nation-building'. This should have led them to be more prudent in the case of Iraq. More generally, history shows that democracy cannot be brought from outside, and that a great many conditions need to be met for a democracy to be a success. Above all, like Hassner, Fukuyama criticizes Krauthammer and those with similar opinions for neglecting the question of legitimacy, which has had devastating consequences for the United States' image around the world. The opinion of others is important, even for the most powerful nation in the world, and Europe's arguments against the war in Iraq were generally just. Contrary to what Krauthammer (and Kagan) has claimed, Europe's opposition to the war was not motivated by the desire to reduce America's freedom to act, nor by a visceral attachment to multilateralism and international law. Europe's refusal to support the United States' position reflects a greater awareness of history and its lessons, and the prudence this engenders. The false arguments put forward by the United States to show the need for the war created a legitimacy gap that forms a major obstacle to America's ability to influence the international scene. The ideas of a mission and the supposed 'exceptionalism' of the United States that are so dear to Krauthammer and others lead them to think that their actions are automatically guaranteed legitimacy, but for the rest of the world this is not at all the case. Only a reasonable and moderate foreign policy could have won the support of other states. Therefore, Fukuyama also insists that the power of the United States has limits and that these limits have been tightened by this lack of legitimacy. In a paradoxical way, the hegemony of the United States and the unipolarity of the international system requires America to act with even greater restraint and to show greater wisdom in order to avoid provoking other states to react to this domination. Consequently, the right to act pre-emptively (carry out preventative attacks) and unilaterally should be used sparingly; it should not be constantly invoked, as it is by Krauthammer. As Ferry pointed out, it also means that the United States must act within a multilateral framework whenever possible and that it must reinforce regional, multilateral organizations.

The debate between Krauthammer and Fukuyama, and the resulting schism in the neo-conservative movement, is an excellent example of the tendency for divisions to form in American schools of thought on foreign policy.[17] The arguments being put forward by moderate neo-conservatives are remarkably similar to those advanced by Jean-Marc Ferry and Pierre Hassner, although there is no evidence that these French philosophers have had a direct influence on moderate neo-conservative thinking. However, what is clear is that Fukuyama, perhaps because of his origins, has been able to distance himself from the collective subconscious and embrace points of view that are often foreign to American culture. It would therefore appear that national differences can be overcome and that a Euro-American consensus can be found.

Notes

1 Kagan's original article was published in *Policy Review* in June 2002. The book *Paradise and Power* was published in 2003.
2 Jean-Marc Ferry, *La question de l'État Européen* (Paris: Gallimard, 2000).
3 See interview with Muriel Ruol, *La Revue Nouvelle*, January–February 2004, <www.fundp.ac.be/interfaces/recherche/recherche_florence.htm> and 'À propos de *La puissance et la faiblesse* de Robert Kagan. Les États-Unis et l'Europe, ou le choc de deux universalismes', in *États-Unis/Europe. Des modéles en miroir*, ed. Jean-marie Ruiz and Mokhtar Ben Barka (Lille, France: Presses du Septentrion, 2006).
4 Ferry, interview with Muriel Ruol, ibid.
5 Ibid.
6 See Pierre Hassner: 'Puissance et légitimité', *Commentaire*, 100, Winter 2002–3, and 'Questions de puissance. Les États-Unis et l'Europe ont-ils deux conceptions opposées des relations internationales?' in *États-Unis/Europe. Des modéles en miroir.*
7 '*Puissance et légitimité*', p. 785.
8 Ibid., p. 788.
9 Ibid., p. 787.
10 Ibid., p. 787.
11 See 'The GOP Won the Cold War? Ridiculous', *The New York Times*, October 28 1992, and an interview published in *New York Review of Books*, August 12 1999.
12 Walter Mead, *Special Providence: American Foreign Policy and How it Changed the World* (New York: Alfred A. Knopf, 2001); Michael Lind, *Made in Texas: George W. Bush and the Takeover of American* Politics (New York: Basic Books, 2003). Kennan's 2002 interview was published in Congress's journal, *The Hill*, September 25 2002.
13 In his interview in *The Hill* (September 25 2002), Kennan flays military adventurism and suggests that the methods and doctrine used to contain Stalin would also have worked for Saddam Hussein.
14 See, in particular, issues 76 (summer 2004), 77 (fall 2004) and 78 (winter 2004/5) of this journal.
15 C. Krauthammer, 'The Unipolar Moment', *Foreign Affairs*, 70, 1990.
16 Francis Fukuyama, 'The Neoconservative Moment', *The National Interest*, 76 (summer 2004).
17 Moderate neo-conservatives, such as Krauthammer, no longer contribute to *The National Interest*. They have formed their own review, *The American Interest*.

5

GERMAN TRANSATLANTICISM

Between narcissism and nostalgia[1]

Wade Jacoby

Germany was in the thick of the transatlantic tensions that erupted in 2002–3. After having pledged 'unconditional solidarity' to the United States in the wake of September 11, the SPD–Green government had to survive a no-confidence vote in order to send 2,500 German troops to fight in Afghanistan. Thousands more stayed on after the war in order to provide security in Kabul. As late as May 2002, George W. Bush could be heard to say, 'We've got a reliable friend and ally in Germany. This is a confident country, led by a confident man [SPD Chancellor Gerhard Schröder].'[2] But as the United States took increasingly strident positions towards the Iraqi leadership over the summer of 2002, the German leadership grew less willing and less able to justify American policies to the left wing of its party and to public opinion. By August 2002, Schröder was referring to US policy as a 'reckless adventure' even as Dick Cheney gave a major speech calling for 'regime change', which the US administration seemed ready to pursue with or without a UN mandate.[3] Fighting an uphill battle for re-election, SPD criticism of the US became increasingly blunt and may have provided the margin of victory in an election that the left won by a mere 7,000 votes. Subsequently, Schröder announced in January that Germany – then a non-permanent member of the Security Council – would not support any UN resolution that legitimised war against Iraq.

In the wake of these events, Germany's image in the United States has taken a beating in many quarters. While US conservatives save the real vitriol for France, many still harbour a fierce grudge against Germany for its refusal to join the coalition of the willing against Iraq. Meanwhile, many Americans who saw wisdom in Chancellor Gerhard Schröder's rejection of US policy, still couldn't miss how ineffective Germany was in influencing the US. What good is being right, they ask, if you are too weak to do anything about it? Schröder's Social Democrats, who ruled in a coalition with the Greens from 1998 until October 2005, answer that their policy spared Germans the misery that American, British, Australian, Polish,

Spanish and Italian soldiers have endured in Iraq. Despite overtures from both sides, anger still simmers between American neoconservatives and the German left.

One of this volume's larger purposes is to give some sense of whether the recent transatlantic tensions are, in any sense, likely to endure. There is no shortage of arguments that deep structural reasons – most importantly, the end of the Cold War – make repairing that break extraordinarily difficult and even unlikely. This chapter approaches the question not from the perspective of deeper political structures but by synthesising the available political strategies as outlined by the two main parties and by a variety of German pundits and commentators who follow transatlantic relations very closely. I encapsulate these strategies under two less than flattering subheadings, narcissism and nostalgia. That the September 18 2005 elections delivered a vote so close that it necessitated a 'Grand Coalition' between the Christian Democrats and Social Democrats only heightens the irony. For some time at least we can anticipate that Germany's foreign policy muddle will be the confluence of the muddle of its constituent parts.

The chapter begins, however, by noting some striking similarities between US neoconservatives and the German left. This step is useful because Atlanticists must understand the parties who generated the split in the first place if they are to have any hope of healing that split. The chapter's equation of the German left with the American right will sound strange to people who have become used to thinking of Europe and the US as Venus and Mars. Aren't the neoconservatives hell-bent on remaking the world one Middle Eastern autocracy at a time, while Germany's most prominent contribution to the 'war on terror' so far is 'city building' in Kabul, where Germany has stationed several thousand troops? Yet there are four key foreign policy similarities between these two very different camps.

First, each government narrowly won its re-election (Red–Green in 2002 and Bush in 2004) while calling central attention to its foreign policy strategy. Thereafter, each group has been inclined towards the dogmatic assuredness that what was best for it electorally was best for its country and, indeed, the wider world. Each is, in that sense, highly narcissistic, even as it talks incessantly of helping others. This self-confidence goes well beyond the garden-variety narcissism of politicians in that each insists that its foreign policy is an essential precondition for global progress on democratisation (Republicans) or world peace (Red–Green).

Second, each group has discovered that the country it governs has major gaps between the ambitious ends of its policy and the more modest means it has at its disposal. The US neoconservatives have discovered that massive military superiority has limited fungibility for post-war tasks, while the German left talks about civilian power often without being in a position to exercise it unless countries with great military power first provide order.

Third, each group insists that states should not join coalitions unless they are 'willing'. This breaks decisively with the German experience in NATO, where its commitments to allies were meant to be automatic in case of any attack. Since the Iraq conflict clearly did not involve such an attack, both parties have shown ample willingness to act unilaterally while calling on the other to act multilaterally. SPD officials invoke long lists of familiar unilateral US actions, but it was they who pre-emptively stated in January 2003 that they would not join a military action against Iraq *even if sanctioned by the UN*.

Finally, both have assumed full freedom to try to assemble *ad hoc* coalitions in ways not overly respectful of the foreign policy history of their two states or their own moral claims. Thus, neoconservatives make new pacts with autocracies like Uzbekistan in the name of democracy promotion, while the German left cosies up to Russia, to the obvious discomfort of Central Europe.

In the face of this constellation, the Christian Democrats have had few good options. Elected in 2005, Chancellor Angela Merkel, is (like Schröder was in 1998) short on foreign policy experience, though her early foreign policy in office has gone smoothly. Currently, however, much CDU foreign policy thinking consists of nostalgic odes to a bygone era when the US and Germany worked arm in arm to deter potential Soviet aggression, and manage the peaceful dissolution of communism and the reunification of Germany. This nostalgia seems misplaced. The behaviour of the German left and the US neoconservatives over the last three years makes the CDU's hopes to go back to the future seem implausible. Where mutual moderation once made US–German frictions manageable, the actions of the recent past have now made them far less manageable, while eroding key coordinating institutions, such as NATO. And nowhere in Europe have opinions of America fallen as far as they have in Germany. Christian Democrats in government may turn both trends – although the Grand Coalition's Foreign Minister is a Social Democrat – but there is no quick way back to the old levels of mutual respect and institutionalised cooperation. How did German transatlanticism come to be stuck between narcissism and nostalgia? What other options are available?

Germany's dilemma

German foreign policy is marked by substantial continuity since World War II. Yet much of Germany's foreign policy dilemma right now is that it is a status quo power in a time when the US is not. A primary reason that Germany is a status quo power is that it is enormously dependent on the current relatively open international economic order. In both 2003 and 2004, the value of its total exports has been larger than that of any other country in the world, including that of the United States, whose economy is over three times as large. Germany still has major economic worries, but no one disputes that they would be much worse if Germany's export performance deteriorated further. Thus any German initiatives to promote a positive change in the wider world – and such initiatives emanate from all German parties – must be reconciled with a strong economic interest in system stability. And even though the end of the Cold War (and the return of full German sovereignty) arguably gave German foreign policy more room for manoeuvre, it has been notably cautious in trying to exercise that room.

Famously, stability has not been high on the US agenda in recent years. Where the Clinton administration made cautious and tentative steps towards promoting democratisation, the Bush team has tried to make the world anew. Challenged by Islamic terrorists and no longer damned by the Cold War standoff to cynicism and pragmatism in human rights, it has had both the motive and the opportunity to go on the offensive. US dynamism, in turn, created new tensions that threatened Germany's room for manoeuvre which had grown since the Cold War's end. Germany's significant fiscal difficulties only compound this dilemma because they make Germany somewhat less able to pacify (let alone buy off) the US by supporting expensive foreign missions. Germany has its own tradition of democracy promotion – as we shall see later – but it works in pluralistic form and at a snail's pace. In short, Germany doesn't have a lot of good options in the face of US assertiveness. Neither its traditional tools nor its short-lived increased room for manoeuvre have been readily available in recent years.

The major German parties have followed two different courses in response to this dilemma. The SPD has trumpeted a moralism that, as suggested above, is sometimes nothing more than narcissism: a tendency to read its own political interests as universal and to ignore the fact that the good that it actually does is not keeping up with the good that it imagines it does. In short, the SPD now exudes a sense that what is good for them electorally is both democratically legitimate and morally just. The CDU, by contrast, has turned to nostalgia for the old transatlantic alliance: a hope that by returning to power in time, they can restore NATO to its proper role, reconnect the frayed lines to Washington, and reassume a traditional role as a mediator inside Europe and across the Atlantic. In order to evaluate these two claims – neither overly flattering – we must first provide quick answers to two very broad questions: What has changed under Red–Green? What can a CDU-led government do differently?

The Red–Green innovations

Although Germany is a status quo power, the Red–Green government has arguably broken with some important German transatlantic traditions.

What are these breaks? The left's detractors emphasize three claims. First, as noted, the Red–Green government actively manoeuvred to block key US policies in Iraq. No past US–German disagreement – and there were many – ever reached this level of public acrimony. Second, at least according to some critics, the Red–Green government has also promoted the rise of anti-Americanism among the German public.[4] Third, it has, also in this more critical view, neglected both NATO and the military aspects of foreign policy that would allow Germany to play a more central role inside NATO, the UN and the Organization for Security and Co-operation in Europe (OSCE).[5]

Red–Green supporters point to a busy legislative agenda. The Red–Green government introduced a host of important reforms in two separate 'security packages' (Sicherheitspaket I and II), which together included about 100 new laws. The first set of laws limited religious freedom for 'extremist' groups and simplified procedures for holding German-based supporters of foreign terrorist organisations. The second set of laws increased the powers of the German Federal Criminal Police and the Border Police, and made easier the process of expelling suspected radicals from Germany. A further law in June 2002 was designed to attack money laundering, and in June 2004 a new law allowed the shooting down of an airplane that was being used as a weapon. At the end of 2004, a new Terror Prevention Office was set up in Berlin with about 180 employees.

Yet the Red–Green government was not prepared to go as far as the US PATRIOT Act, and the Grand Coalition faces many open questions to confront. In the wake of the London bombings in July 2005, the CDU has called for a constitutional change that would allow the *Bundeswehr* to be used in domestic terror situations. Since the change would require a two-thirds majority, SPD support is essential. Representatives of the police union have called for four further steps, most of which have also been endorsed by the CDU: (1) more personnel to observe suspects (in the summer of 2005, there were approximately 160 active investigations linked to Islamic terror suspects); (2) the introduction of a comprehensive anti-terror database; (3) more flexible rules regarding conducting and storing telephone taps; and (4) improvements in emergency services.

Having noted these major policy shifts and the accompanying legislative agenda, a substantial dose of narcissism still registers in the left's foreign policy. By increasingly identifying SPD electoral interests with German national interests and even with universal interests, the SPD basks in the warm glow of pursuing a 'policy of peace in the national interest'.[6] The formula has something both for the German left – though only part of it is pacifist – and for the more centrist voter, for whom 'national interests' can be an attractive slogan. Moreover, at a time when the SPD was imposing domestic economic reforms that were deeply controversial inside the party, 'being right' on Iraq was an enormous strength for the Chancellor. Earlier in his tenure, Schröder took some large risks – on both Kosovo and Afghanistan. Each of them could have brought down his government – points that are barely noticed in the US debate about Germany. But this was during a time when his party still hoped that its economic reforms would spark more growth. With his reputation as an economic manager badly tattered by unemployment levels around 5 million and a series of disappointing labour market reforms, Schröder relied heavily on the legitimacy gained at home from standing against the US.

Thus, the more pragmatic left of 1999–2001 swung back toward a longing for large-scale peaceful civilisational projects such as combating world hunger or AIDS. But longing is not enough. True, Germany stations plenty of troops abroad, but those troops are not highly capable, and if they often help prevent order from unravelling, they seem to make relatively little contribution toward restoring order. Moreover, throughout the second Red–Green term, the Green Foreign Minister – who continued to articulate reasonably well the German dilemma brought by major moral ambitions but modest resources – conspicuously lost ground to the Chancellor, who seemed more concerned with emphasising German self-confidence rather than Germany's limited ability to achieve the ambitious ends his government articulated.

CDU aspirations

Meanwhile, the German centre-right is palpably nostalgic for its role as a quiet and steady American partner inside a NATO that is the centre of Western political-military strategy. That this NATO no longer exists is occasionally noted but usually as a prelude for arguing that Germany needs to do whatever it can to bring it back. As with the SPD, the US policy shift contributed to the party's dilemma: when the US rejected a moderate and multilateral policy, the Red–Green government was hard pressed to provide cooperation as they had in Bosnia, Kosovo, Macedonia and Afghanistan. By losing interest in NATO as a military organisation, the US hollowed out the institution that anchors all CDU conceptions of foreign policy.[7]

One reason nostalgia is so appealing, therefore, is that Germany doesn't have a lot of other good options. There are a few highly demanding alternatives, including Adam Posen's claim that Germany should partially redeploy its assets from the European level to the global level and play the role of global 'economic hegemon' that the US has recently left unfilled.[8] There are parts of this agenda that the CDU could potentially warm to. But nostalgia is both cheaper and easier and, therefore, it appears more realistic in a government that seems unlikely to devote many resources to foreign policy. The question is whether the nostalgic road will lead anywhere.

To answer this, we first have to ask whether the CDU intends to restore the policies – especially the US policy blockade, anti-Americanism, and

NATO issues noted above – with which many felt Red–Green had broken. On the first issue, the CDU will clearly try to position the German government close to the US, rather than trying to block it. How to do so is a major topic discussed below. On the second issue, the open criticism of the US, the new government will choose between risking an opening to the US, which might inflame public opinion, and being a quieter partner. The Green and Left opposition and even parts of the SPD may be quick to imply that the CDU would make Germany into a US 'vassal'. The CDU understands the potential electoral damage this theme can do them, and while the Grand Coalition, a new election might be required at any moment. On the third issue, the new government will miss no opportunity to call for the strengthening of NATO. Yet it is less clear that the US will have any interest in heeding this call. This is especially true since the German government is likely to have almost no room for manoeuvre on the fiscal side and will, therefore, be obliged to argue that their increased contributions to NATO will come from better efficiencies with existing resources.

All of these items were featured in the 2005 CDU election manifesto, which also called for closer relations with the US and more distance from France and Russia. In the wake of the London bombings, the CDU called immediately for a substantial regional presence of the *Bundeswehr*. Their central idea is to 'bundle' civil and military capacities in each area and provide 'joint leadership' (*gemeinsame Führung*) in case of emergency. Part of this involved opposing SPD efforts to cut reserves forces, but the CDU also called for a move back to more territorial defence (which positions them to use the military on their own territory in case of terror attack). It strongly emphasised a reinvigoration of NATO.

Three versions of the new Atlanticism

It is notoriously difficult to read a party's foreign policy strategy from its campaign promises, and all the more so when it enters government with its primary political rival. For reasons just noted, the CDU is constrained in its ability to seek public *rapprochement* with the United States. Therefore, to get deeper into the possibilities now open in Germany, we need to look into the thinking of the Atlanticist community within Germany. Many of these thinkers are politically inclined toward the CDU, as evidenced by affiliation with the party itself or by activities under the aegis of the party's Konrad Adenauer Foundation.[9] I group these 'new Atlanticist' responses under three categories: nostalgia, resignation, and embrace. The first involves some form of reconstitution of traditionally friendly US–German relations. The second sees the past value of Atlanticism but is resigned to the need for Germany to manage a very large and enduring gap between Germany and the US (and partially Britain). The third camp sees a foundation for a new Atlanticism in close partnership with the US and Britain by essentially embracing the policies of the Bush administration.

Nostalgia

The bulk of CDU-friendly intellectuals support some variant of the policy of nostalgia identified above, though many of them have thought much more deeply about the roots of the current dilemma than have CDU officials. Much of their writings are deeply grounded in the history of German foreign policy. For example, Christian Hacke argues that the Red–Green government abandoned Germany's traditional foreign policy role as a modest mediator between the US, UK and France and became instead the noisy, hasty junior partner of France.[9] While France preserved its room for manoeuvre until very late in the Iraq crisis, the Red–Green government committed itself categorically (and unilaterally) against the war in January 2003. Worse, by abandoning its traditional mediating position, Germany complicated the calculations of all other states. While France ultimately chose Germany's side, Britain and Central and Eastern Europe had little choice but to swing much closer to the United States than might otherwise have been the case. As Hacke puts it, 'the politics of division that Germany accuses the US of was something it practiced itself'.[10] For this camp, Germany's policies made transatlantic polarisation worse than it would otherwise have been. Noting a long list of previous German–US disagreements, Hacke concludes that, 'until 1989, Germany had a provincial character but a professional foreign policy. Now it is nominally sovereign, but its foreign policy is amateurish.'[11] At bottom, Hacke's central point is that German interests lie with America, so while it can disagree with America quietly and firmly, it must never be in open opposition to American policy. German interests cannot be pursued with a 'small European' vehicle.[12]

Karl-Heinz Kamp is another CDU-friendly writer, though more critical of the US and somewhat less critical of Red–Green.[13] Kamp emphasises Germany as a mediator[14] not so much between the different national states (like Hacke) as between the US and the EU. The US knew how much weight Germany had and used its close relations with Germany to help influence the shape of the EU. If German weight inside the EU lags – and Kamp thinks that is already happening – the partnership is then worth much less to the US. Kamp notes that the CDU was 'vague' about Iraq (and that the US noticed this as well). He says Germany's hasty decision put it in a very awkward position inside NATO, which came to the fore on the Turkey issue, where Greece, to take the most striking contrast, was ready to defend Turkey, but Germany was not. Still, Kamp argues that prophets of transatlantic doom have been wrong before and that the transatlantic relationship can be re-established on the basis of Germany committing more fundamentally to NATO, resisting any Common Foreign and Security Policy (CFSP) that may counter the US, engaging in a military reform that spends more money on fewer tasks, and engaging in public diplomacy to raise the profile of foreign policy issues among the German public. The CDU's dilemma, however, is that they have no money to raise defence spending, and no strong desire to play up an issue that is still perceived as one of the SPD's strengths. Any nostalgia here will have to be both cheap and relatively quiet.

Few German voices have been as consistently Atlanticist as Arnulf Baring's. Baring's recent writing has two main themes: first, the US is Germany's essential partner and its only alternative to foreign policy 'isolation'. Second, Red–Green has mistakenly imagined that it could stand against the US and limit US options, only to discover that it is itself limited and isolated. Baring wants to see Germany escape this isolation and return to close ties with the US. Baring praises the policies of the first Red–Green government, noting especially that neither the Kosovo conflict nor the sending of German troops to Afghanistan would have been easy for a CDU-led government. But he is furious at what he sees as the unnecessary confrontation with the US over Iraq. Baring claims in particular that Schröder promised Bush twice that he would not actively oppose US policy on Iraq. He also argues – against the grain of much other CDU-friendly writing – that the East German floods meant that Schröder didn't need the confrontation with the US to win the election. Baring concludes that the carefully wrought German image has been destroyed by Red–Green 'dilettantes'.

Most nostalgist criticisms of Red–Green leave important matters unexplained. Most importantly, they often imply that Germany had a chance to stay out of the war – which they knew then to be highly unpopular and know now to be badly run – without paying the high diplomatic costs of an open break with the US. But the nostalgists don't convincingly show how that was possible. Though they relentlessly criticise the Red–Green government for leaving Germany with little room for manoeuvre, it is not clear how much room Germany would have had in its traditional '*Vermittler*' position. For example, Hacke admires the professionalism (if not the policy) of the French government. Unlike Germany, at least the French kept their cards covered until the last hour. Yet if Germany had not announced early on that it had heard enough evidence to conclude that the coming war was a flawed and possibly illegal undertaking, then would France have been able to play its cards as it did? Would it not have been possible – even likely – that both Germany and France would, by waiting to the last moment, have run out of good options and joined the US position? Certainly, this is exactly what happened in very many states (including many in Eastern Europe) and what the Bush administration clearly hoped would happen with France and Germany. Having participated, however grudgingly, in the 'coalition of the willing' would make Germany a different place today. If these authors would have preferred that, they should say so more clearly. In particular, they should probably cast the same critical eye on Edmund Stoiber's pronouncements that they cast on Schröder's. Those who imply that Germany should have joined with the US are often silent on the democracy costs of governments that act against the overwhelming preferences of their populations. One could argue that German public opinion would have been somewhat more supportive of the war had the Red–Green government also supported it, but that seems true only to a very limited extent. Government positions varied strongly across Europe, but public opinion varied a lot less. One could also argue that foreign policy is no place for democracy, but again, that should be argued out rather than left implicit.

Finally, nostalgists often make much of Central and Eastern European (CEE) support for US policy toward Iraq. In their telling, CEE elites, despite their misgivings, saw loyalty to the US as a long-term interest that outweighed their own electoral fortunes. This rather flattering picture of CEE elites misses the 'thinness' of CEE governments' commitment to US policy toward Iraq. The CEE states did not, in fact, simply ignore their pacifist publics. They also signed on to the US-led initiative in highly ambivalent ways, including having an outgoing president (not the government) sign the letter in the Czech Republic and failing to follow proper civil-military procedures in Hungary.[15]

Resignation

Other Atlanticists are resigned to the idea that the close US–German relationship – wonderful while it lasted – will likely stay broken for quite some time. But if they are resigned to that essential fact, they are far from resigned to the idea that Germany is without options. For example, Werner Link has been a leading Atlanticist intellectual for many years, but he now openly brands the US an imperial hegemon.[16] His central concern is that the US has so aggressively challenged the already-weakened Westphalian system that it may turn 'enduring freedom' into 'enduring war'. Like Baring and Hacke, he emphasises Germany's traditional role ('*Spagatpolitik*'), but unlike them he downplays the influence of short-term SPD electoral considerations. Rather, Link sees German foreign policy as aligning itself longer term with a 'balancing' coalition of France, Russia and China. He argues that the balancing coalition is not likely to be overtly hostile to the US, but rather to 'cooperatively balance' US power.[17] Neither is each balancing coalition likely to be permanently fixed. Rather, despite his grave misgivings about moving away from the US, it is clear that Link sees the balance as a kind of leavening and moderating influence, a check and balance, rather than some kind of veto.

To pursue such a policy, Link thinks German elites will need a thicker skin. He takes issue with his Atlanticist friends in both Germany and America by denouncing their tendency to cry 'anti-Americanism' at every turn and to use false claims of a purported German anti-Americanism as a 'club' to hammer at any objections to American policy. As capable as the nostalgists of drawing on history, Link notes that many important French–German initiatives – from the Élysée Treaty, to the German–French brigade, to the EuroCorps – were pushed through despite American objections. He also downplays the fear of East Europeans about independent German foreign policy, noting that East Europeans understand that a Germany embedded in the EU is not the Germany they feared in the past. Nevertheless, he sees Europe caught in a 'precarious situation' having rejected American hegemony without yet having a 'Europe-puissance' to replace it with.

Like the nostalgists, this camp sees some hope in complementary roles between the US and Germany. Where Baring, Hacke and American writers like Charles Kupchan worry that the US may distance itself much more significantly from Europe, Richard Herzinger calls the US and Europe 'Siamese twins, with admittedly different bone structures'. Herzinger underscores the strangeness of formulations of 'America versus Europe' at a time when several European states supported US policy. In the wake of the 2004 US election, one could add that tens of millions of US voters do not support Bush's policy. For Herzinger, the West is characterised less by agreement on what it wants than by agreement on what it doesn't want: repression and dictatorship. Thus, rather than the old system of paying lip service to consensus or the hopelessly naïve new idea of a European counterweight, Herzinger suggests that we are now in an era where the 'negative consensus' that holds the West together will have to be complemented by a difficult process of openly articulating different perspectives and coming to a series of *modi vivendi*.[18]

The great disadvantage is that this camp seems to have little sense of how a US–German *rapprochement* could be established. The nostalgists seem to imply that Germany should swallow its pride and return to support of America's main line while quietly disagreeing at the margins. This camp sees that concession as going too far but then seems limited to hoping for a change of heart in Washington before staking out an agenda for restoring cooperative transatlantic relations. To the extent that important elements of American unilateralism preceded the Bush administration and are quite likely to survive it, this is a considerable omission.

Embrace

Besides the wistful resignation to the fact that Germany and the US have parted ways, there is another non-nostalgic view from the right. Put simply, the American neoconservatives are right, and Germany ought to join them under CDU leadership. For example, Joachim Krause has essentially taken on the task of translating neoconservatism into German. Krause identifies four sources of order: raw balance of power, rule-and-value-based realist arrangements, institutional guarantees, and the spread of democracy and markets. All have coherent, if different, conceptions of what order is and how it can best be maintained. The German debate is more likely than the American debate to emphasise that the Cold War was waged successfully through a combination of balancing Soviet power and embedding Soviet power in deals that made its power more transparent, and through efforts to promote liberal values to Soviet-dominated peoples.[19] Krause argues that Germans make the mistake of collapsing the debate about international order into a simple condemnation of American unilateralism and thereby eliding factors like the 'zone of liberal order' that encompasses great swathes of Europe and Asia, the fragile but real legitimacy of the United Nations, and the dynamism of globalisation that challenges traditional state tools for providing order.

Krause then defends the American neoconservatives as having the courage of their liberal convictions: they are trying to stabilize and expand the existing zone of freedom, even leaving intact institutions that operate inside it (e.g. WTO). But outside that zone, they are aggressive and don't hew to the conservatism of classical realism, but rather seek to eliminate the sources of problems. They recognise, says Krause, that globalisation brings with it the chance to spread liberal practices much more quickly and efficiently than ever before. Krause calls this a 'pragmatic' liberalism because, unlike earlier versions of liberalism, it does not pretend its values are universal. But it does seek to push them as far as it can and then play hardball outside that sphere. Krause argues that Europe, by contrast, has had no serious foreign policy debate for fifteen years. And while Germany pursues a liberal conception of order with some institutionalism on the side, France has a classic balance of power conception with institutionalist window dressing. The strains between Germany and France are thus held to be even greater than were those between the Bush Administration and Red–Green.

If the US and Germany are both pursuing liberal concepts of order, where does the tension come from? Krause blames Germany. First, he sees German liberalism as 'universalistic' rather than pragmatic: it is uncomfortable exercising power but also not prepared to compromise. Second, where US institutionalism sees the UN as a potential tool for identifying and combating states, German institutionalism highlights procedural issues of sovereignty over the 'substantive goals' of the UN Charter. For Krause, Germany worries about what is legal more than it worries about what is right. Third, Germans worry less about terrorism and weapons of mass destruction and more about environmental issues than do US liberals. These latter problems require multilateral solutions, which crucially often include deals with non-democratic states. Still, on balance, Krause sees plenty of common ground between liberals on both sides of the Atlantic, especially if the moralism of the Germans could give way to more pragmatism about liberalism's ends.

Ulrich Speck's approach to German interests is different.[20] First, he argues that the SPD–Green government talked big talk in areas like human rights and north–south development, but essentially continued CDU–FDP policies in these areas. Second, he argues that German foreign policy must always attend to the need for multilateral institutions that can secure an orderly international economy while reassuring Germany's partners that it is not a danger to them. Germany has a comfortable position as a second-tier actor in these military campaigns – a role that doesn't require it to have an overall strategy. A *Handelsstaat* also doesn't care about the domestic conditions of its trading partners, as long as they stick to contracts. Speck also criticises the inadequacy of a foreign policy dominated exclusively by economic concerns and, like Hacke, he ridicules Europeans for having no foreign policy concept towards the Middle East. He characterises European policy to promote order in the Middle East and in Africa as 'colonialism without responsibility' in which Europeans buy raw materials from corrupt rentier regimes that repress their own people. The cost of this strategy is increasing, and the dangers it generates are increasing as well. As such, Speck represents an alternative to nostalgia because he sees the need for a European transformation strategy for the Middle East, something Europe has never had and to which it logically cannot return.

There are important silences in the embrace camp. First, Krause and Speck make no comment about the workability of bringing democracy to Iraq and the Greater Middle East, which seems the least we could ask of a self-styled pragmatist such as Krause. Second, they take at face value that the Bush administration has a coherent grand strategy, apparently on the grounds that it has published one. But grand strategies are notoriously in need of interpretation. For example, what could a German government have known about US intentions in the spring of 2003? Those who say Germany should have stuck with the US must also argue that Iraq was a one-off event, not the first step in a long war to remake the Middle East. This claim is easier to make in 2005–2006, with American troops bogged down in Iraq. But no one could have known that in 2003, and many signs pointed then to the opposite, with rhetoric like 'you're next' aimed at Iran and Syria. Third, while German critiques of Red–Green fault the government for 'feeding' anti-American sentiment, these authors have little to say about the way Washington has nurtured and channelled the understandable anxieties of Americans into a robust electoral strategy. While the Bush administration surely hoped for a more successful course for the Iraq war and lost some votes because of its handling of the war, the far more important point is that by successfully connecting the Iraq war to a war on terror, the Bush administration mobilised key constituencies around its call for more time to finish the job. In particular, many of the things Krause dislikes most about the Red–Green government – the lack of a practical long-run plan, the ubiquity of partisan politics, the relentless use of scapegoats – are everyday occurrences in Washington. His telling grants the Bush foreign policy a coherence it lacks while denying it an electoral strategy it enjoys in abundance.

Atlanticist amnesias

For all the reasons noted above, it is a tough time to be an Atlanticist. German–American relations will not get back on track until the contours of both US and German foreign policy are more firmly established. Germany needs a clear conception of its foreign interests. The US needs to decide if it is keeping the foreign policy it has adopted in recent years. Until those two fundamental choices are made, Atlanticists can take only very partial steps towards reconciliation. But even modest steps are made harder by some characteristic ‘amnesias’ indulged in by Atlanticists of all stripes. Moreover, if these points are often forgotten by those who value the US–German relationship, they are often *terra incognita* for those who do not, such as many of the US neoconservatives and those elements of the German left mentioned in the introduction. This chapter concludes with three points that American and German Atlanticists, respectively, are prone to leave out of their analyses.

First, American Atlanticists forget too easily that the growing German assertiveness in military terms in the first years of the Red–Green government (e.g. Kosovo, Macedonia, Afghanistan) was premised on the NATO umbrella. With NATO slipping into military irrelevance in the eyes of the US, it was very likely that a Germany led by the SPD would become far more cautious. There is no doubt that this trend was hastened by an alarming lack of consultation. Germany, France, the US and Britain all missed lots of chances to consult in advance of policy shifts. But the bigger issue is that there is no widely esteemed forum right now for managing these problems. Here, it seems that constructing new transatlantic institutions is beside the point. The problem is the lack of sustained, high-level political interest in consultations and policy coordination. New institutions will not solve that problem, and if the interest returns, NATO’s North Atlantic Council will suffice. A Grand Coalition may well try to raise the profile of NATO, but it will not be successful if the US continues to regard NATO so lightly.

Second, US Atlanticists consistently forget or undervalue the contribution to stability made by EU enlargement. Partly, this is because US officials are focused on their own demands – especially the intense operations tempo in the US military – and get angry when they see other states with different priorities. At one level, US State Department officials, even in the Bush administration, are quick to acknowledge the importance of EU efforts to promote stability in Central and Eastern Europe. Occasionally, as in the Orange Revolution in Ukraine, it even seems like old times with the US and the Europeans pulling more or less on the same lines (and echoes the good efforts to synchronise support to Serbian Non-Governmental Organizations (NGOs) and opposition parties in the late 1990s). Yet these same officials are quick to talk of a European 'usability gap' in addition to the long-lamented 'capability gap'. Even where the Europeans have assets, they are reluctant to deploy them, say these officials. This is certainly accurate as far as it goes, but again it misses the fairly intense engagement that Germany has had with Central and Eastern Europe in past years and the positive contribution that it has made there. That said, the Grand Coalition may limit German support for future EU enlargement, and without the carrot of membership, we should not expect the EU to play the same positive role in reforming states.

Third, American Atlanticists frustrated with Germany must remember that moving on to a war in Iraq fitted German interests *much less well* than did the initial attack against Afghanistan. It was easier to make a case that Afghanistan posed direct threats to German interests because the Taleban clearly harboured terrorist killers whose fatwahs often lumped all the West in one camp. This was not true of Iraq, at least not in 2003. US Atlanticists must remember that the domestic justification for war will likely always be difficult to achieve in Germany and that it will generally be *harder* with the right in power than with the left. Thus far, German governments have no experience fighting wars in the teeth of a strong parliamentary opposition. Of course, a Grand Coalition could only go to war with the support of both parties, but one might then see on one or both sides the kind of 'backbench' rebellions that have plagued the Labour government in Britain.

German Atlanticists tend to suffer three different amnesias. First, some of the more impatient Atlanticists tend to forget that the US won't be ruled by neoconservatives for ever. It would take a long time – or a serious terrorist attack by Islamic radicals – to change German public opinion on this issue, and there seems little chance that the CDU will spend a lot of political capital to try (though both major parties may be well prepared to do so should a terror attack come on their watch). Absent that, by the time Germany could move enough to cooperate closely with this White House, America may be led by a different kind of leader. More broadly, German Atlanticists should remember that the size of the gap in US–German relations is determined more in Washington than it is in Berlin simply because Berlin has less room to move.

Second, German Atlanticists should remember that Germany too has interesting tools for promoting democracy. Some of these tools are bilateral, including the world's best-funded party foundations, which work with a variety of different non-governmental actors inside foreign countries. Others are multilateral and often coordinated through the EU or OSCE. Both forms of organisation are less susceptible to partisan capture and more insulated from the vagaries of public opinion than the corresponding American instruments. The German party foundations were, after all, the largest foundations to try to affect state actions in CEE. Of course, the party foundations' shaping ability is helpful in the long term, but hard to depend on in the short term. The party foundations can't easily steer individual states, and Germany is much more successful when working with the EU to shape neighbouring states.[21] Neither tool is viable in areas of major civil unrest, but each is a modest and useful tool that German Atlanticists can point to with some pride.

Third, as noted earlier, German Atlanticists often elide the point that on Iraq Germany probably really couldn't be certain of staying out of the conflict militarily while still staying close to Washington. If Germany had not taken an early and clear stance when it did, France would have had much less time and much less assurance in preparing for the massive break with the US that occurred. Used to modelling moderation, Germany suddenly modelled a high stakes gamble. This point is important because many nostalgists dislike Bush's style quite intensely but see the partnership with the US as essential. They tend to argue as if Germany could have achieved the substantive policy its population wanted – no combat participation – without paying such high costs *vis-à-vis* Washington. They think Schröder overpaid for peace, but was a lower price really available? Precisely because Schröder lacked self-confidence and experience he may well have needed to play his cards quickly in order to have any chance of securing the substantive policy he (and the overwhelming majority of Germans, including a lot of nostalgists) actually wanted: to sit out the war against Saddam Hussein. German Atlanticists should not forget how hard the choices really were in their eagerness to damn Schröder.

Notes

1 The author thanks Christian Hacke, Patrick Keller, Michael Krekel and Jana Puglierin for helpful comments on an earlier draft.
2 Quoted in Stephen Szabo, *Parting Ways: The Crisis in German–American Relations* (Washington, DC: Brookings, 2004), p. 19.
3 Ibid., pp. 19–33.
4 Joachim Krause, 'Multilaterale Ordnung oder Hegemonie?' *Aus Politik und Zeitgeschichte*, 31–2, 2003, pp. 6–14.
5 Christian Hacke, 'Was bleibt von der rot–grünen Aussenpolitik', *Neue Züricher Zeitung*, 15 Juni 2005, 6ff.
6 See the SPD election manifesto at www.spd.de.
7 Wade Jacoby, 'Military Competence Versus Policy Loyalty: Central Europe and Transatlantic Relations', in David Andrews (ed.) *The Atlantic Alliance Under Stress: US–European Relations After Iraq* (New York: Cambridge University Press, 2005), pp. 232–55.

8 Adam Posen, 'If America Won't Germany Must: A Globalizing World Needs a New Economic Hegemon', *Internationale Politik – Transatlantic Edition*, Summer 2005, pp. 32–7.
9 None of this should deny that there are strong Atlanticists in other parties. However, Atlanticists friendliest to the CDU are likely to be the best clue to the range of options open to the party.
10 Christian Hacke, 'Deutschland, Europa und der Irakkonflikt', *Aus Politik und Zeitgeschichte*, 24–5, 2003, pp. 8–16.
11 Ibid., p. 11.
12 Ibid., p. 14.
13 Hacke presciently notes that the revolt of the Atlanticists inside European states may make European cooperation much more difficult: ibid., p. 16.
14 Karl-Heinz Kamp, 'Die Zukunft der deutsch amerikanischen Sicherheitspartnerschaft', *Aus Politik und Zeitgeschichte*, 46, 2003, pp. 16–22.
15 For details, see Jacoby, 'Military Competence'.
16 Werner Link, 'Imperialer oder pluralistischer Frieden', *Internationale Politik*, 5, 2003, pp. 48–56.
17 Ibid., p. 52.
18 Richard Herzinger, 'The American and European Siamese Twins', *Internationale Politik* (International edition), 14, 3, 2003, pp. 39–42. See also Moïsi, 'Die Weiderfindung des Westens', *Internationale Politik*, 12, 2003, pp. 21–44, and Thomas Risse, 'Es gibt kein Alternative: US und EU müssen ihre Beziehungen neu justieren', *Internationale Politik*, 6, 2003, pp. 9–18.
19 Krause, 'Multilaterale Ordnung oder Hegemonie?' pp. 3–4.
20 Ulrich Speck, 'Deutsche Interessen. Eine Kritik der rot–grünen Außenpolitik', *Merkur*, February 1 2004, pp. 106–16.
21 See Christoph Bertram, 'Zentral, nich bloß normal', *Die Zeit*, 32, 2004.

6

BRITAIN, THE UNITED STATES AND EUROPE

To choose or not to choose?

John Baylis

> As a leading European power she will speak with great authority to the US and her influence in Europe is likely to depend both on her own strength, military and economic, and on the extent to which she is known to enjoy influence and support in the US.[1]

This statement by Foreign Office officials could easily have been written in recent years to describe the aims of British diplomacy towards Europe and the United States. Both Prime Minister Tony Blair and former Foreign Secretary Jack Straw have gone out of their way in recent years to emphasise Britain's pivotal role in acting as a 'bridge' between the two sides of the Atlantic, explaining and interpreting each to the other. With the serious transatlantic differences over the Iraq war this role has been perceived in Whitehall as being of critical importance. The fact that these words were written by officials not in 2004 or 2005, but in March 1949, demonstrates a consistent line in British diplomacy, reflecting a determination by successive governments to play a distinctive role designed to enhance Britain's influence in both the United States and Europe. This chapter sets out to argue, however, that this role has been significantly complicated, both in the past and at present, by the policy priority in Whitehall in favour of the 'special relationship'. Despite the rhetoric that Britain doesn't have to choose between Europe and the United States, it is clear that in some important respects Britain did choose as a result of the traumatic experience of the Second World War and that this legacy has continued (with brief exceptions) to guide British diplomacy and statecraft ever since. To develop this argument I begin by looking at the legacy of the Second World War.

The impact of the Second World War on the development of the 'special relationship'

What is striking about contemporary British diplomacy, especially towards the United States, is the continuity with the past, especially the importance of the symbols, memories and experienced relationships associated with the Second World War. Anglo-American relations obviously have deep roots going back to 1776, but, as H.G. Nicholas and other writers have shown, it was in the struggle against the Axis powers that the community of the English-speaking peoples was tested most and British leaders, some more reluctantly than others, came to recognise that close ties with the United States were critical to the nation's survival and continuing well-being.[2]

It was in the period before Pearl Harbor, that, despite American neutrality, a 'common-law alliance' was established between Britain and the United States. During this period 'a gradual mixing-up process' took place, with wide-ranging cooperation involving such things as lend-lease arrangements, the destroyers-for-bases agreement, and the beginnings of intelligence collaboration. All of these arrangements helped to lay the foundations of the 'full-marriage' which was to follow.[3] Once the United States had entered the war in December 1941, the formation of a joint war machine quickly developed. This began with the 'Arcadia' conference in January 1942 and the confirmation of the 'Germany-first' strategy, followed by the formation of a series of Combined Boards, which together played a crucial role in the direction and coordination of the allied war-effort. The war against Germany, as well as against Japan, and the two vital areas of Anglo-American cooperation in the fields of atomic energy and intelligence, together highlighted the remarkable degree of collaboration that was achieved. This was reflected in General George C. Marshall's far-reaching claim in 1945 that the Anglo-American partnership during the war represented 'the most complete unification of military effort ever achieved by two allied states' in the history of warfare.[4]

Of vital importance to the successful wartime alliance (and the concept of the 'special relationship' which developed from it) was the close personal relationship established between Winston Churchill and Franklin Roosevelt. The nature of this relationship can be summed up in the following (perhaps apocryphal) story. Shortly after the Japanese attack on Pearl Harbor, Churchill arrived in the United States for discussions with the American President. During his visit, the Prime Minister stayed in the White House. According to the historian Robert Sherwood, on one occasion, Roosevelt was wheeled into Churchill's room only to find him emerging from the bath, 'wet, glowing and completely naked'. Somewhat disconcerted, Roosevelt started to leave the room only to be called back by his guest. 'The Prime Minister of Great Britain', he declared, 'has nothing to conceal from the President of the United States.'[5]

True or not, this story illustrates some of the main features of what came to be known as the 'special relationship'. In particular, it reflects the importance of the personal relationship between the leaders of both countries and also the informality, as well as the mutual trust which is said to characterise the relationship in general.[6] Such qualities, together with a common language and common culture, are often seen as setting the Anglo-American relationship apart from other 'normal' inter-state relationships in international politics. Churchill even went as far as to suggest joint citizenship between the two countries in 1943. This proposed combined identity would involve a common currency and a common trading area. The attempt to bind Britain and the United States ever closer together was part of a deliberate strategy by Churchill. He was increasingly conscious during the war of the growing disparity of power and Britain's dependence on the US. He later recalled: 'No lover ever studied every whim of his mistress as I did those of President Roosevelt.'[7]

The concept of a 'special relationship' between Britain and the United States was initiated by the repeated and emphatic use of the term by Churchill in his Fulton Speech on 5 March 1946. In this speech the former Prime Minister spoke about the vital importance of the wartime partnership which resulted from the 'fraternal association', 'the growing friendship' and the 'mutual understanding' between the two 'kindred systems of society'. This rhetoric was taken up by post-war historians such as Sir Dennis Brogan, Arthur Campbell Turner and H.C. Allen who pointed to the 'linguistic and cultural relationship between England and America' which set it apart from the relationship which either state had with any of its other allies.[8] This emphasis on shared history, culture and language is also a feature of some American interpretations of the 'special relationship'. George Ball, a former Under-Secretary of State, pointed out that:

> to an exceptional degree we look out on the world through similarly refracted mental spectacles. We speak variant patois of Shakespeare and Norman Mailer, our institutions spring from the same instincts and traditions and we share the same heritage of law and custom, philosophy and pragmatic Weltanschauung . . . starting from similar premises in the same intellectual tradition, we recognise common allusions, share many common prejudices, and can commune on a basis of confidence.[9]

In some important respects, the idea that 'sentiment' as much as 'interest' is at the root of the 'special relationship' which these accounts portray has been the dominant one in the historiography of Anglo-American relations. A study of the archival evidence, however, suggests a rather different view.[10] While it has long been known that close Anglo-American relations were a more-or-less continuous objective of British foreign policy for much of the twentieth century, there has been less awareness of the fact that British governments both during and after the Second World War attempted consciously to use the concept of the 'special relationship' to reinforce British interests. Close ties with the US had been, and were likely to be in the future, vital to Britain's survival and therefore had to be given priority.

On 21 March 1944 a major Foreign Office paper was written entitled 'The Essentials of an American Policy'. After commenting on 'the special quality' of the Anglo-American relationship which had developed in the war, the paper emphasised the crucial need for skilful British diplomacy to influence the direction of US foreign policy in the future.

> They have enormous power, but it is the power of the reservoir behind the dam which may overflow uselessly, or be run through pipes to drive turbines. The transmutation of this power into useful forms, and its direction into advantageous channels, is our concern.[11]

It was argued in the paper that after the isolationism of the past the Americans were thinking for the first time about playing a more direct role in world affairs. In this context, it was believed that it was important not to pursue what had been the traditional policy of balancing British power against that of America, but instead to make use of American power for purposes which suited the British government. It was suggested that the key objective of British diplomacy was to steer 'the great unwieldy barge' that is the US in the right direction.

> If we go about our business in the right way we can help to steer this great unwieldy barge, the United States of America, into the right harbour. If we don't, it is likely to continue to wallow in the ocean, an isolated menace to navigation.[12]

This emphasis on superior British wisdom in directing the less experienced United States in the right direction, particularly away from isolationism, was to become a recurrent theme in the years that followed. As such, the concept of a 'special relationship' was a deliberate British creation, or what David Reynolds has described as 'a tradition invented as a tool of diplomacy'.[13] It was an important part of a policy that reflected not only a perceived superiority emanating from long diplomatic experience but also a belief that Britain remained a world actor with a responsibility for helping to maintain international peace and stability.

Post-war debates about a 'Third Power' role

In the immediate post-war period there was initially a rapid cooling of the close wartime relationship. This was highlighted by the break-up of the integrated war machine in 1945–6, with the abrupt cancellation by the United States of the lend-lease arrangements, the winding up of the Combined Wartime Boards and the unilateral ending of nuclear cooperation by the United States, despite wartime agreements that it should continue. There were also suspicions in the United States of British socialism, which were reflected in the difficulties which arose in post-war negotiations over a loan for Britain to see her through the immediate economic crisis bequeathed by the war. Britain regarded the terms offered by the United States as unnecessarily hard, reflecting a failure by the Truman administration to recognise the effort the British people had put into the war, initially alone.

The fact that the 'unwieldy barge' was difficult to guide in the right direction was an important part of the decision by the Labour government in January 1947 to develop an independent nuclear deterrent. As the Prime Minister Clement Attlee was to explain later, because of traditional US isolation, Britain couldn't always rely on American governments in a crisis:

> We had to hold up our position vis-à-vis the Americans. We couldn't allow ourselves to be wholly in their hands, and their position wasn't awfully clear always. At that time we had to bear in mind that there was always the possibility of their withdrawing and becoming isolationist once again.[14]

In March 1949, shortly before the North Atlantic Treaty was signed, a major review was undertaken in the Foreign Office of 'Anglo-American Relations: Present and Future'. Britain, it was argued, had a choice between a 'Third Power' grouping or a close partnership with the United States.[15] It was a choice, the paper said, which should not be dictated by 'ties of common feelings and tradition' but by 'a cold estimate of advantage'. It was decided that the 'Third Power' concept should be rejected because 'the partnership with the United States is essential to our security'. The best policy for Britain, it was suggested, was to be closely related to the US, but independent enough to influence American policy. This would require Britain to remain a major European and world power and to sustain its own independent military capabilities. Britain must be the partner, not the poor relation, of the United States.[16]

This reflected the British view in the late 1940s that there was no significant contradiction between a European policy and a 'special relationship' with the United States. Indeed the two were complementary. Playing a leading role in Europe would reinforce British influence with the US and close ties with the US would strengthen Britain's position on the continent. As Manderson-Jones has shown, this was also a view held by the Truman administration in the United States. Both governments supported the idea of 'a "three-pillar" Atlantic community, with Britain as a master-link between the United States and Continental Europe'.[17] The war had shown that Britain needed to maintain US involvement in European security, but it had also shown that Britain could not, as it had in the past, turn its back on Europe. Its destiny was tied up with developments on the continent. Britain's key role in the formation of the Western Union and NATO reflected this dual approach to statecraft.

Cold War and the defence priority

During the early 1950s, however, it became increasingly clear that there was tension between the two key strategic objectives of British foreign and defence policy. The proposals put forward by France for a European Defence Community created a dilemma for the Conservative government of the day. As the Cold War gathered momentum, Britain had a strong interest in seeing closer defence collaboration between the Western European states. At the same time, there was little interest in Whitehall in submerging British armed forces in a supranational defence organisation. Britain was (perceived to be) an independent great power with world-wide interests and a close partnership with the United States. When asked to choose, Britain declined to join the new defence identity, thereby helping to ensure its ultimate failure.

The EDC proposals, together with the Messina Talks and the Treaty of Rome, highlighted the fact that the continental states were moving in a direction that the British disapproved of, but over which they had little control. The 'special relationship' with the United States provided an alternative policy which would reinforce Britain's great power aspirations, ensure her security in the context of Soviet nuclear weapons and conventional preponderance in Europe, and provide backing for the pursuit of her global interests. To achieve this policy, it was believed that it was important to demonstrate to the United States that Britain was a worthy partner. Bevin had denied in 1947 that Britain had 'ceased to be a great power'.[18] In line with this, the strategy adopted in the 1950s was designed to show that Britain remained a reliable partner of the US and a power that was 'vital to the peace of the world'. This was one of the main reasons for Britain's support for the United States during the Korean War, despite the serious long-term effects which this had on the British economy. The transformation of Empire into Commonwealth also reflected Britain's determination to continue to play a world role. Britain's self-image as a global power, at this time, was not unreasonable, even if it did breed some illusions. Britain was still the world's third major state in the 1940s and 1950s – economically, militarily and in nuclear capability. As such, it remained a valuable ally for the United States.

In order to maintain her great power status and importance to the United States, Britain exploded her first atomic weapon in October 1952 and took the decision in July 1954 to develop thermonuclear weapons to match those first tested by the United States and the Soviet Union in 1952 and

1953 respectively. One of the interesting features of de-classified documents on Anglo-American relations is that despite the frequent rhetorical emphasis of the Churchill government on the 'fraternal association' of the two states, in the early 1950s, one of the main objectives of British policy was to seek influence over US policy making. This was necessary because there was considerable anxiety in government circles about the failure of the United States to provide Britain with information about nuclear planning even though the presence of American bases in Britain (granted in 1948) made it a major nuclear target for the Soviet Union in the event of war. There was also considerable concern that American rashness and impatience might precipitate a global conflagration. Even the Prime Minister himself was concerned about the worrying trends in US policies. In May 1954 he warned Cabinet colleagues of American impatience:

> I know their people – they may get in a rage and say . . . why should we not go it alone? Why wait until Russia overtakes us? They could go to the Kremlin and say: These are our demands. Our fellows have been alerted. You must agree or we shall attack you.[19]

From the perspective of the government in the mid-1950s it was imperative for Britain to develop thermonuclear weapons, not only to maintain 'her rightful place as a world power', but also to restrain the United States. This reflected a widespread belief in Britain that it would be dangerous if the US were to retain their present monopoly 'since Britain would be denied any right to influence her policy in the use of this weapon'.[20] A British H-bomb was seen as a device to open up nuclear cooperation with the United States which had been ended in 1946 and to allow Britain to influence and guide American policy in a more responsible direction. Once again we see the belief in Britain's greater wisdom and the importance of using the 'special relationship', especially in the military field, in order to achieve greater influence over American policy making.

The Suez Crisis of 1956 was a traumatic shock to Britain's great power pretensions and to its policy of pursuing a 'special relationship' with the United States. Faced with American opposition, the Eden government was forced to back down in humiliating fashion, demonstrating the reality of British dependence on, rather than influence over, the United States. For the French, the lesson of Suez was to seek greater independence and a greater role in Europe. For the British, however, the lesson was that there was a need to develop even closer interdependence with the United States and to avoid the growing integration that was taking place on the continent at the time.

As Jan Mellisen and others have shown, from 1957 to 1963, the Macmillan government put considerable effort into re-creating an intimate and harmonious relationship with the United States.[21] Reflecting past policy, the Prime Minister believed that Britain could 'play Greece to America's Rome, civilizing and guiding the immature giant'.[22] This was particularly important, in Macmillan's view, in the nuclear field. At the Bermuda and Washington Conferences in 1957, he sought to build on his close wartime friendship with Eisenhower to create a climate of trust at the highest level of government. The renewal of the 'special relationship' which resulted was reflected in the deployment of Thor missiles in Britain, joint strategic planning between the air forces of both countries and the achievement of Britain's long-term objective of repealing the 1946 McMahan Act, prohibiting full nuclear co-operation between the two countries. As a result of the last, and most important, of these agreements, Britain was to receive preferential treatment in the form of information from the United States on the design and production of nuclear warheads, as well as fissile material. This set the scene for the intimate nuclear relationship which was to follow. When Britain cancelled its Blue Streak intermediate range missile in 1960 on grounds of cost and technological obsolescence, the United States stepped in to offer the air-launched Skybolt missile for use with the RAF's V-Bomber force. And when the Kennedy government cancelled Skybolt in late 1962, the Polaris submarine-launched missile was made available to Britain at the Nassau Conference in December at a bargain-basement price.

Nuclear politics at Nassau

The Nassau Conference in December 1962 provides a good illustration of the way the British were prepared to manipulate the concept of the 'special relationship' to achieve their objectives. After the cancellation of the Skybolt missile in 1962 which Britain had agreed to buy from the US to maintain the nuclear deterrent, there were those in the State Department who saw the opportunity to get Britain out of the nuclear business. American support for the British nuclear deterrent was seen as an impediment to the pursuit of a non-proliferation policy and US interests in a more united Europe. At the beginning of the conference, Kennedy appeared reluctant to provide Britain with Polaris missiles which the British government regarded as the only suitable substitute for Skybolt. Faced with the possibility of a major diplomatic defeat, Macmillan, as he later admitted, had to pull 'out all the stops'. In a highly emotional speech, he referred back to the halcyon days of the Second World War and the foundation of the 'special relationship' between the two countries. If agreement could not be reached now, after all these years, he argued, then he would prefer not to patch up a compromise. 'Let us part as friends . . . if there is to be a parting, let it be done with honour and dignity.' Britain would not welch on her agreements (by implication, as the Americans had done). Switching tack, he then went on to ask the American President if he wished to be responsible for the fall of his government. He warned that if this happened there would be a wave of anti-American feeling in Britain and even the possibility that an anti-American faction might assume the leadership of the Tory party in an attempt to cling on to power. The result would be the end of the close and harmonious relationship between the two countries. In response to this eloquent and evocative appeal, Kennedy gave in and Britain got the Polaris missiles it wanted.[23]

George Ball, one of the State Department officials advising Kennedy, later argued that the President had been seduced at Nassau by 'the emotional baggage' of the 'special relationship' which, in his view, had got in the way of cooler judgement. Nassau, he believed, was an illustration of how the United States 'had yielded to the temptations' of a myth. He argued that:

> US interests in both a strong and united Europe and the prevention of nuclear proliferation had been harmed by the over-zealous support for the partnership with Britain, especially in the defence field.[24]

He also suggested, however, that the British themselves had become victims of their own rhetoric. Because they had come to believe in the 'special relationship', they had failed to adjust their foreign and defence policies to the reality of their reduced status in the world. The close ties with the US had encouraged successive governments, Ball argued, 'in the belief that she could by her own efforts' play an 'independent great power role and thus it deflected her from coming to terms with her European destiny'.[25]

Attempts to re-balance Europe and the 'special relationship'

Even when British governments in 1963 and 1967 did attempt to readjust their foreign policy by applying to join the Common Market, the concept of a 'special relationship' between Britain and the United Sates proved to be an impediment. This time it was the use made of the concept by Charles de Gaulle in vetoing the British applications to join the European Community. Whether he believed it or not, the French President rejected British membership on the grounds that Britain remained an Anglo-Saxon power, wedded to close ties with the United States. Britain, he argued, was not a truly European power. When things went wrong, either economically or militarily, Britain would always revert to the Anglo-Saxon partnership.

While the nuclear and intelligence links between Britain and the United States remained in the 1960s and 1970s, faced with de Gaulle's opposition, there was a concerted attempt by the Wilson and Heath governments to play down the 'special relationship'. In symbolic fashion, the Wilson government deliberately talked of a 'close', rather than a 'special', relationship between the two countries.[26] In reality, the relationship also became less special as a result of Britain's withdrawal from east of Suez, eroding the global partnership which had been such an important aspect of Anglo-American relations in the 1940s and 1950s. Britain also refused to participate in the Vietnam War, which further undermined an already tetchy personal relationship between Wilson and Johnson.

Heath made a similar attempt to symbolise the changing nature of British policy in the early 1970s by referring to the 'natural relationship', rather than the 'special relationship'. As a committed European, the Conservative Prime Minister quite deliberately turned his back on many of the conventions which had characterised the 'special relationship' in the past. Whether Heath's rejection of the concept of the 'special relationship' was crucial in Britain's successful application to join the European Community in 1973 is doubtful. Nevertheless, it did provide a symbolic indication of a desire to change the direction of British foreign policy and to re-invent a British European identity.

The 'special relationship' restored

The priority given to Europe over America, which characterised Heath's policies, was not followed, however, by his successors. In the late 1970s the Labour Prime Minister James Callaghan sought to maintain Britain's position within the EC, but at the same time, he fostered close personal links with the Ford and Carter administrations in the United States. It was left to Margaret Thatcher, however, in the 1980s, to attempt to resurrect the 'special relationship' after 'the rope had been allowed to go slack' in the 1960s and 1970s. The close personal relationship between Thatcher and Reagan laid the foundations for what the Prime Minister described as the 'extra-ordinary relationship'.[27] Her abrasive and confrontational approach towards the European Community contrasted with her determination to play the role of the most loyal of America's allies.

The restoration of the 'special relationship' during the Thatcher era was symbolised by the Trident agreement of July 1980 and March 1982 and the close military (and intelligence) collaboration during the Falklands War. Despite her misgivings, the Prime Minister was also prepared to face considerable domestic political costs in supporting the deployment of Cruise missiles in Britain, not being too critical of American actions in Grenada, and allowing American F111 aircraft to use British bases to bomb Libya in 1986. In the last case, as Margaret Thatcher points out in her memoirs, there were significant benefits from supporting the United States. Special weight, she says, was given to British views on arms control negotiations with the Russians and the administration promised to give extra support to the extradition treaty which the government regarded as vital in bringing IRA terrorists back to Britain.[28] The fact that so few had stuck by America in her time of trial, she suggests, strengthened the 'special relationship'.

It is clear from Mrs Thatcher's discussion of the Libyan crisis in her memoirs and her treatment of Anglo-US relations in general that 'interests' rather than 'sentiment' were the crucial basis of the 'special relationship' as far as the Prime Minister was concerned. This is revealed in her comment, 'I knew that the cost to Britain of not backing American action was unthinkable.'[29] The continuity going back to the Korean War is clear. Whether she deliberately used the 'special relationship' as a tool of diplomacy during her period in office to reassert Britain's place on the world stage will only become known when the archives are opened. It seems likely, however, that she saw the 'special relationship', like many of her predecessors, as a useful device to harness American power in the pursuit of British interests. It is also probable that the Prime Minister believed in Britain's superior wisdom in guiding (Ronald Reagan and) the United States in the right direction during the crucial events which took place in the 1980s. In terms of 'culture and sentiment', the 'special relationship' symbolised her perception of Britain as an important Anglo-Saxon state, different from, and in some ways superior to, her European allies.

Uncertainties of the post-Cold War era

With the end of the Cold War and the disappearance of Margaret Thatcher and Ronald Reagan from the political scene, the 'special relationship' initially underwent a serious challenge. Claims that the Major government interfered in the election that brought President Clinton to office, difficulties over the Bosnia peace plans, and differences over Sinn Fein President Gerry Adams's visits to the United States in the mid-1990s encouraged a number of commentators to argue that Anglo-American relations were 'special no more'.[30] Even those who were sympathetic to close ties between Britain and the US, like former American ambassador Raymond Seitz, argued that it was 'likely that our priorities won't match with quite the same frequency as they once did, and the overlap of our strategic interests may not be quite as extensive as before'.[31]

The election of the Blair government in 1997, however, led to a renewal of close ties. Clinton and Blair developed a good working relationship and shared an ideological commitment to 'third way' political ideals. The American President also played a very constructive role in the Northern Ireland peace process and joint military operations in northern Iraq and Kosovo highlighted the value placed on the continuing diplomatic and defence relationship between the two countries. Despite the close personal ties between the two leaders, however, there were also some policy differences. Even during the joint air campaign in Kosovo the Blair government found it difficult to persuade the Clinton administration that a ground campaign might be necessary to bring about the final withdrawal of Milosovic's forces. There were also growing unilateralist trends in US foreign policy that caused increasing concern for Britain and other European states. As David Ryan has argued: 'Europeans collectively approved, where the United States did not, the International Criminal Court (ICC), the 1997 Kyoto Protocol on climate, the ban on land mines, the bio-diversity treaty, and the mechanism for the verification of the Biological Weapons Control Treaty.'[32]

These concerns were exacerbated, at least initially, by the election of George W. Bush as President of the United States in November 2000. In particular, there was anxiety that the loss of the close personal and ideological ties at the highest level would inevitably lead to an erosion of the 'special relationship'. Worries over American opposition to the new 'autonomous' European Rapid Reaction Force and apparent British concerns over even greater unilateralist tendencies in the United States raised question marks over the general convergence of Anglo-American interests which had taken place during the Blair/Clinton era. Blair's visit to Washington in February 2001, however, indicated that contemporary obituaries of the 'special relationship' were premature. Blair put a lot of diplomatic effort into reassuring the new American President not only that the new European force would not undermine the NATO Alliance but also that the United States could expect British support if it went ahead with its plans to develop a missile defence shield. At the same time, Bush pleased his visitors by using the term 'special relationship' to describe the continuing relationship with Britain. The visit highlighted the determination of both governments (and especially the British) to retain the close ties of the past.

The new global crisis

That determination was demonstrated even more dramatically in the aftermath of the attacks on the Twin Towers and the Pentagon in the autumn of 2001. Two things were noticeable about the reaction of the Blair government in the immediate aftermath of the attacks on September 11: first, the constant message of standing 'shoulder to shoulder' with the American government and the immediate offer of support, and second, the deliberate search to build a broad international coalition within hours of the atrocity occurring. The conventions of the past were immediately on display. Britain had to be seen to be the most supportive (and toughest) of America's allies. It was also important that Britain should be perceived to be prepared to aid the US in a practical and major way in its hour of need, including military aid in Afghanistan to topple the Taleban government in October 2001. British officials argued that this provided the greatest opportunity to influence the Bush administration in the dangerous period ahead.[33]

Much the same was true of Britain's involvement alongside the United States in the Iraq war in March 2003. Despite the debate over the legitimacy of the pre-emptive attack on Saddam Hussein's Iraq and the widespread condemnation both domestically and amongst the international community, the Blair government was prepared to take considerable risks in support of its American ally. Although the British government was supported by some of the newer recruits to the EU, it was prepared to go against the firmly held views of the French and the Germans, thereby putting the futures of the UN, NATO and even the EU itself at risk. Apart from the fact that the government believed that this was the 'right' thing to do, ministers went out of their way during the conflict to indicate that at a time of growing unilateralism in the US support for the Bush administration in the war was the only effective way of trying to influence American policies in an increasingly unipolar world.

The British role in the Iraq war clearly reflects a basic continuity in post-Second World War British foreign and defence policy. Amongst the decision-making elite there has been a more or less consistent and enduring belief that Britain's ultimate security and economic well-being depends on close ties with the United States.[34] The perceived 'lessons' of the 1939–45 war, of the British way of life being saved by American intervention, continued to be reinforced during the Cold War by a belief that close ties with the US were critical in balancing the overwhelming power of the Soviet Union. September 11 and the toppling of Saddam Hussein both appear to have confirmed the contemporary importance of this traditional policy. During the war against Iraq, in particular, great efforts were put into portraying the similarities between the Bush and Blair relationship and that of Churchill and Roosevelt. Referring to one of the wartime meetings between British and US leaders in March 2003, one commentator suggested that 'the choreography of the Camp David war council, so reminiscent of the FDR–Churchill meetings on that very spot, seemed to echo the greatest moments of the Anglo-American Alliance'.[35]

Whether this continuity with the past is wholly positive, however, continues to be debated in Britain. The historian Sir Michael Howard has argued that the experience of the Second World War (involving the close relationship with the United States) has had an unfortunate influence on British foreign and defence policy down to the present day. Although it was Britain's 'finest hour', he argued, the nation 'has continued to re-live it disastrously'. In Howard's view, Britain's

> position at the top table in NATO gave it a sense of superiority over its continental allies – countries which it had either conquered or liberated – who sat below the salt. The significance of their economic recovery was under-rated. Their plans for the creation of a European Community were treated with contempt. The special relationship with the United States prolonged British delusions of grandeur. It took the humiliation of Suez to bring home to the British the reality of their position in the world and to force belated re-adjustments.[36]

For opponents of the 'special relationship' the 'readjustments' have still not fully taken place. The concept of the 'special relationship' undoubtedly continues to have a powerful effect on British thinking, as the Iraq war has demonstrated. It is often suggested that the concept has what one US official has described as a 'misty quality'. Critics argue that it summons up, consciously or unconsciously, deep-seated feelings that Britain remains an important power on the world stage ('punching above its weight'); that it should maintain wide-ranging international responsibilities; and that it has an Anglo-Saxon identity, superior to, and in some important ways separate from, its European partners. As such, it is often argued that the continuing use of the term tends to encourage a nostalgia for the past at a time when Britain needs to be thinking more independently and creatively about its future role in the EU in the aftermath of the war in Iraq.[37]

The former British Ambassador to the United States, Robin Renwick, wrote in 1996 that:

> Britain will continue to wrestle with its European destiny. For reasons of politics, history and geography the country will never be at the heart of Europe, in the way France, Germany and the Benelux countries are. . . . The British have made a commitment to Europe, but it will remain peculiarly allergic to being governed from outside their borders. They will never be in the forefront of European integration. Hence the paradox that Britain cannot afford to be marginalised in Europe if it is to remain influential in Washington; yet in some important respects it is precisely the fact that Britain *is* different to other European countries, and has been willing to act without waiting for a European consensus, that has rendered the relationship valuable to the United States and effective in action.[38]

Conclusion

Whether the US will continue to see Britain's role in Europe in the future as of major importance is less certain. Senior American officials have indicated that traditional US interest in dealing with Europe as a single unit might well be replaced by closer bilateral ties with those who have been, and are, prepared to support American interests. This raises question marks over Britain's traditional attempts to act as a bridge between the two sides of the Atlantic. As the events associated with the Iraq war have shown, Britain continues to feel closer to the United States than it does to Europe. Both leading political parties also continue to see the 'special relationship' as an important instrument of foreign and, especially, security policy. Given the overwhelming contemporary preponderance of American power, supporters argue that it makes sense to maintain the 'special relationship' as the key part of British foreign and defence policy.

As a result of the Iraq war there are those who argue that Britain has finally chosen the US over Europe and that this represents a 'profound' change in direction of British policy after decades of trying to act as the bridge between Europe and the United States.[39] In practice, however, as this chapter tries to show, the 'bridge' concept has (almost) always been based on a British priority in favour of the US side of the bridge. It is true that there have been considerable efforts by British ministers since the war in Iraq to emphasise Britain's continuing commitment to the EU. Tony Blair and Jack Straw have both gone out of their way to argue (for good political reasons) that Britain would continue to refuse to make a choice between Europe and the United States. Given the range of continuing interests Britain has in Europe, and the importance British governments have given to close UK/EU/US relations in the past, there has been a determination to ensure that, on the big issues, Britain is at the heart of the debate. It is also true that on some issues, like Iran and Kyoto, Britain sometimes sides with Europe rather than the US. *Not* choosing is seen by the government as being in British interests.

Nevertheless, the Iraq war reinforced the lesson that when there is a serious crisis Britain always sides with the United States. It has also highlighted the continuing tension between trying to balance the European and Atlantic strands in British diplomacy. The former British Ambassador to the United States, Sir Christopher Meyer, highlighted this tension in an interview in November 2005 when he argued that: 'you don't have to choose between America and Europe, but it is difficult to ride both'.[40] The contemporary dilemma for Britain is whether it should try to lead what Donald Rumsfeld called 'the new Europe' in developing a new transatlantic relationship, independent of France and Germany, or whether it should seek to try to heal the frictions between 'old Europe' and the US and re-establish the traditional European–US relationship? Given the structural changes which have taken place in international relations in recent years, Britain is attempting to undertake two delicate diplomatic tasks. One is to steer the 'great unwieldy barge' back towards closer relations with 'old Europe'. The other is to act as the bridge not only between Europe and the US in general but also between 'new' and 'old' Europe. No doubt Britain will continue to pursue both strands of policy rhetorically, refusing to choose, in the hope of persuading its European allies to support its preferred Anglo-Saxon model for Europe. Such a model, however, reflects the continuity of relative choice in favour of maintaining close ties with the United States. The contemporary disagreements between Britain, on the one hand, and the French, on the other hand, over the future of Europe, however, suggests that the objectives of this statecraft may well be just as difficult to achieve in the future as they have been in the past.

Notes

1 Public Record Office, PRO: FO371/76384, 'Third World Power or Western Predominance', 23 March 1949.
2 See H.G. Nicholas, *The United States and Britain* (Chicago: Chicago University Press, 1975); A.P. Dobson, *Anglo-American Relations in the Twentieth Century* (London: Routledge, 1995); and C.J. Bartlett, *The Special Relationship: A Political History of Anglo-American Relations since 1945* (London: Longman, 1992).
3 See J. Baylis, *Anglo-American Defence Relations, 1939–1984: The Special Relationship* (London: Macmillan, 1984).
4 G.C. Marshall, *The Winning of the War in Europe and the Pacific,* Biennial Report of the Chief of Staff of the US Army, 1 July 1943 to 30 June 1945, to the Secretary of War.
5 Robert Sherwood, *Roosevelt and Hopkins: An Intimate History* (New York: Harper Brothers, 1948), p. 442.
6 See H. Kissinger, *The White House Years* (New York: Weidenfeld and Nicolson, 1979), pp. 90–1.
7 R. Renwick, *Fighting with Allies: America and Britain in Peace and War* (London: Macmillan, 1996), p. 271.
8 See D. Brogan, *American Aspects* (New York: Harper and Row, 1964); A.C. Turner, *The Unique Partnership, Britain and the United States* (New York: Pegasus, 1971); H.C. Allen, *The Anglo-American Predicament* (London: Macmillan, 1960), and *Great Britain and the United States: A History of Anglo-American Relations* (London: Odhams, 1955).
9 G. Ball, *The Discipline of Power* (London: Bodley Head, 1968), p. 91.
10 See J. Baylis, *Anglo-American Relations since 1939: The Enduring Alliance* (Manchester: Manchester University Press, 1997).
11 PRO: FO371/38523, 'The Essentials of an American Policy', 21 March 1944.
12 Ibid.
13 D. Reynolds, 'Roosevelt, Churchill and the wartime Anglo-American alliance, 1935–1945: towards a new synthesis', in H. Bull and W.R. Louis, eds., *The 'Special Relationship': Anglo-American relations since 1945* (Oxford: Clarendon, 1986), pp. 85–6.
14 Quoted in F. Williams, *A Prime Minister Remembers* (London: Heinemann, 1961), pp. 118–19.
15 PRO: FO371/76384, 'Third World Power or Western Preponderance', 23 March 1949.
16 Ibid.
17 R.B. Manderson-Jones, *The Special Relationship: Anglo-American Relations and Western European Unity 1947–1956* (London: Weidenfeld and Nicolson, 1972), p. 106.
18 Hansard's Parliamentary Debates, vol. 437, col. 1965 (16 May 1947).
19 Lord Moran, *Churchill: Taken from the Diaries of Lord Moran. The Struggle for Survival 1940–1965* (Boston: Houghton and Mifflin, 1966), p. 580.
20 PRO: DEFE 4/70, Note by the First Sea Lord, 12 May 1954.
21 J. Mellisen, *The Struggle for Nuclear Partnership: Britain, the United States and the Making of an Ambiguous Alliance, 1952–1959* (Groningen: Styx, 1993).
22 Quoted in J. Roper, 'British Perspectives of the United States: Historical and Cultural Bases of the "Special Relationship"', *US Today*, September 1988.
23 PRO: PREM 11/4229, Record of a Meeting held at Bali-Hai, the Bahamas, at 9.50 a.m. on Wednesday, 19 December, 1962.
24 Ball, *The Discipline of Power*, p. 93.

25 Ibid.
26 H. Wilson, *The Labour Government 1964–70* (London: Penguin, 1974), p. 80.
27 See *The Sunday Times*, 23 December 1979.
28 M. Thatcher, *The Downing Street Years* (London: HarperCollins, 1993), p. 449.
29 Ibid., p. 444.
30 See J. Dickie, *'Special' No More* (London: Weidenfeld and Nicolson, 1994).
31 Speech to the Pilgrims Society, London, 19 April 1994.
32 D. Ryan, 'Ten Days in September: The Creeping Irrelevance of Transatlantic Allies', *Journal of Transatlantic Studies*, 1.1, Spring 2003, p. 23.
33 See 'British–US Relations', Foreign Affairs Committee Report, HC 327, HMSO, 11 December 2001.
34 See J. Baylis, *Anglo-American Relations since 1945*.
35 Andrew Sullivan, 'Winds of War are Blowing Britain Away from Europe', *The Sunday Times*, 30 March 2003.
36 See Sir Michael Howard, 'Victory that Exhausted the Nation', *The Times*, 28 May 1994.
37 Concern over the damaging effects of the use of the term 'special relationship' also exists in official circles in Britain. Sir Christopher Meyer, who was British Ambassador to the United States from 1997 to 2003, refused to allow the use of the phrase in the embassy. See Christopher Meyer, *DC Confidential* (London: Weidenfeld and Nicolson, 2005), p. 56.
38 Renwick, *Fighting with Allies*, p. 281.
39 See Sullivan, 'Winds of War are Blowing Britain away from Europe', and Irwin Stelzer, 'So, EU or US, Tony? You Are Going to Have to Choose', *The Times,* 10 March 2003. See also W. Underhill, T. McNicoll and S. Theil, 'Who Speaks for Europe', *Newsweek*, 7 April 2003, and G. Wheatcroft, 'A Relationship that is Now – Your Country Right or Wrong', *Guardian*, 27 January 2003.
40 Interview with Sir Christopher Meyer in *The Sunday Times*, 13 November 2005.

7

THE UNITED STATES AND THE COMMON EUROPEAN SECURITY AND DEFENCE POLICY

No end to drift?

Steve Marsh

The United States (US) was midwife to and Cold War guardian of European integration. The North Atlantic Treaty Organisation (NATO) projected US power and influence into Europe and was the linchpin of transatlantic security co-operation against the Soviet Union (USSR). American influence in Europe was underpinned further until the 1970s by the Bretton Woods system and by the 'special relationship' with Britain, which for reasons of its own dutifully played the role of Washington's loyal lieutenant. The European Union (EU) contributed to Cold War security by overcoming through integration Franco-German enmity and establishing a zone of peace throughout Western Europe. Beyond this, however, potential European alternatives to NATO, notably the European Defence Community in the 1950s and French aspirations to resurrect the Western European Union (WEU) in the 1980s, fell foul variously of national sovereignty concerns, Cold War calculation and American opposition.

The end of the Cold War shook this transatlantic bargain in ways still being felt today. The US was left as an unrivalled hyperpower with strategic interests rapidly shifting from Europe to Asia and the Middle East and with its policymakers uncertain what to do with, and how best to preserve, the 'unipolar moment'. The EU, liberated of strategic dependence on America, anxious to tie-in reunified Germany, and sensing an opportunity to push for deeper political integration on the back of the Single Market programme, sought a stronger security role in the lead up to the Treaty on European Union (TEU). Contemporaneously, NATO descended into an existential crisis upon the dissolution of the Warsaw Pact and the Anglo-American relationship was quickly consigned by some scholars to history as being 'special no more'. Such nay-sayers believed that the loss of a common threat meant that Anglo-American strategic interests would no longer coincide as they had, that deepening British involvement in European integration would weaken Atlanticist ties, and that Britain's self-appointed Atlantic intermediary role would become untenable.

It is instructive of transatlantic relations to analyse against this background US policy towards the EU's post-Cold War evolution as a security actor and, in particular, its development of a Common European Security and Defence Policy (CESDP). This chapter assesses early American calculations of its post-Cold War interests in Europe and argues that the consequent US approach to a European Security and Defence Identity (ESDI) and CESDP has been premised around three key considerations, namely burden sharing, NATO primacy and an informal seat at the EU table. These goals have been pursued with consistency and in partnership with Britain, which has considered its own interests to be best served by binding the US to Europe through a reformed NATO. However, while US control of CESDP development has thus far been relatively successful through direct pressure and by proxy, and while NATO and the EU have begun military co-operation, there are reasons to question whether this situation is sustainable in the medium to long term.

The US and post-Cold War Europe

Referring to the end of the Cold War, Robert Gates, former Director of the Central Intelligence Agency, noted sadly that 'the greatest of American triumphs – a triumph of constancy of purpose and commitment sustained over four decades at staggering cost – became a particularly joyless victory'.[1] Enormous challenges of winning the peace rapidly eclipsed the 'winning' of the Cold War. US policymakers had to define America's place in the post-Cold War world, evolve a replacement for containment strategy, adjust American priorities to new circumstances and make sense of the geo-strategic implications consequent upon the retrenchment of Soviet power. And all of this had to be done at a time when America's ability to bask in the unipolar moment was clouded by pressing domestic issues and by influential predictions of its relative decline.

An immediate consequence of the Cold War's end was to strip Western Europe of its long-standing centrality in US geo-strategic priorities. Gorbachev's renunciation of the Brezhnev Doctrine, signature of the Conventional Forces in Europe Treaty and acceptance of German reunification effectively terminated Western Europe's front-line status. The subsequent dissolution of the Warsaw Pact and of the Soviet Union confirmed this and accelerated the George H. Bush administration's shift of focus toward Asia and the Middle East, especially in the wake of the first Gulf War. Bush's Secretary of State James Baker indicated quickly the extent to which American strategic interests in Europe had diminished when, upon the implosion of Yugoslavia, he declared: 'We don't have a dog in this fight.'[2] Clinton then added to Europe's geo-strategic down-grading through his elevation of geo-economic considerations and his administration's development of the Big Emerging Market programme, within which only Poland of the European economies featured. Clinton's belated commitment to the Balkans would have been unthinkable during the Cold War and eventually owed less to US national security than to secondary considerations for NATO credibility and American leadership of democratic enlargement.

None of this meant that the US either would or could neglect Europe. It still provided the bulk of America's allies and the US was locked into profound economic interdependence with the EU especially. What was needed was a renegotiation of the transatlantic relationship that sustained US influence in Europe but at dramatically reduced cost. This necessitated that NATO find a new *raison d'être* and adapt to post-Cold War conditions. Evidence of this determination came quickly. The North Atlantic Co-operation Council marked the onset of NATO outreach and on 8 November 1991 a New Strategic Doctrine was unveiled that laid claim to responsibilities far beyond collective defence. Partnership for Peace subsequently ratcheted up co-operation and confidence-building measures with non-NATO countries and, with Russia slipping in Washington's concerns, formal NATO enlargements followed in 1999 and 2004. Meantime, the EU had to be persuaded to assume more of the burden for exporting security, especially within its immediate locale. After all, its normative and economic power was a potential stabilising force in Europe, and twin intergovernmental conferences in readiness for the TEU promised improved capacity for purposive security action.

Bush consequently agreed in July 1989 that the European Commission should co-ordinate G24 financial assistance to Central and Eastern Europe and worked hard to establish a greater security as well as economic partnership with the EU. In November 1990 the Transatlantic Declaration formally recognised 'the accelerating process by which the European Community is acquiring its own identity in economic and monetary matters, in foreign policy and in the domain of security'.[3] The New Transatlantic Agenda (NTA) set out this idea of security partnership more robustly in December 1995. According to US Ambassador to the EU Stuart Eizenstat, this was 'the first time we have dealt comprehensively with the EU, not simply as a trade and economic organization, but as a partner in a whole array of foreign policy and diplomatic initiatives'.[4] Under the NTA the US and the EU were, for example, to exercise joint leadership in the consolidation of democracy in Russia and Central Europe and in the reconstruction of Bosnia and Herzegovina and to concert their efforts in preventative diplomacy, provision of humanitarian assistance and promotion of multilateral free trade.

It was important that the US also maintained strong relationships with key European capitals in order to reaffirm American influence and shape the development of integration. Three potentially key relationships were with Moscow, Bonn/Berlin and London. Once the George H. Bush administration lost the USSR as a potential partner in stability, it retained a strategic interest in co-operation with Russia, principally on account of the Soviet nuclear legacy but also to guard against potential reassertion and its promotion of European security structures alternative to NATO. In the West, Bush welcomed Chancellor Kohl and a reunified Germany as a 'partner in leadership' of a New World Order – much to British Prime Minister Thatcher's chagrin. Germany was an established driver of European integration and looked set to be an economic powerhouse in the heart of Europe and a key player in exporting security eastwards. Additional assets were its strong export orientation, which would help guard against a 'Fortress Europe', and its ability as a supporter of NATO to restrain France within the EU, the latter being a long-standing advocate of balancing US influence in Europe. And then there was Britain. Relations initially cooled after the remarkably close Reagan–Thatcher era, especially during the Clinton–Major years, which US Ambassador Raymond Seitz regarded as run on a 'grin-and-bear-it basis'.[5] Nevertheless, Britain remained a staunch Atlanticist and relations improved substantially once Blair became Prime Minister in 1997.

These US reassessments and calculations led it to support broadly the development of an ESDI and to develop a threefold focus centred on guaranteeing NATO primacy, pressing for greater European burden sharing, and maintaining an informal US seat at the EU table. US rhetoric with regard to the first of these has been remarkably consistent. The Dobbins Demarche of February 1991 set out the George H. Bush administration's opposition to WEU development outside of NATO. Secretary of State Madeleine Albright warned in December 1999 against the so-called 'three Ds' of de-linking, duplication and discrimination in the development of an ESDI. Secretary of Defence Rumsfeld reprised similar arguments in February 2001, warning that '[a]ctions that could reduce NATO's effectiveness by confusing duplication or by perturbing the transatlantic link would not be positive' and risked injecting instability into an enormously important Alliance. Even the more diplomatic Secretary of State Colin Powell emphasised in December 2003 that 'the one thing that we feel strongly about is that we should do nothing which would, in any way, put at risk the established structure of NATO or how we do planning within NATO . . .'.[6]

American administrations welcomed increasing EU soft security initiatives but still pushed consistently for greater and more effective European spending on military capabilities and on military-related Research and Development (R&D). European free-riding was unacceptable and their usefulness as allies would be severely compromised unless greater attention and resources were directed towards ensuring interoperability and abandoning obsolete Cold War platforms and troop-heavy forces. This criticism is not unwarranted. Neither is American suspicion that an expanding NATO Europe drains from the US more than it contributes. In 2004 the US dedicated 3.9 per cent of GDP to defence spending, which totalled approximately $462,099 million – equivalent to the next thirty-two most powerful nations combined. That same year NATO Europe collectively spent $235,374 million at an average 1.83 per cent of GDP, well below the world average of 2.6 per cent. Matters within the EU are still worse from the American perspective. In 2001 the Union combined spent just 47.3 per cent of what America did on defence, or in terms of GDP an average 1.75 per cent compared to 3.2 per cent. Moreover, the EU member states' spend to capability ratio is vastly inferior to America's, owing to duplication, protection of national champions in the defence industry, under-investment in R&D and slow adjustment of force structures to post-Cold War military demands. The European Commission estimated in 2003 that the real military capability of the then EU15 was approximately 10 per cent of America's. Germany has been a particular target of American ire. Between 1990 and 2003 its defence expenditure halved from 2.8 per cent of GDP to 1.4 per cent; 2004 spending continued at this rate, German military reforms failed to abolish expensive conscription policies and further spending cuts are likely given high levels of German taxation and public debt.

Increased European military collaboration, improved procurement co-ordination and joint research projects can potentially help compensate partially for relatively low military spend. The US thus welcomed ideas for a strengthened European pillar within NATO and evidence of EU determination to pull its weight better in providing for international security following acceptance of responsibility for the Petersberg Tasks in the Amsterdam Treaty and subsequent Anglo-French agreement in St Malo in December 1998 that the Union should have the hard security capabilities to meet these commitments. Plans for a 60,000-man-strong EU Rapid Reaction Force (EURRF) by 2003, the development of CESDP and agreement of successive Helsinki Headline Goals to develop necessary capabilities were all steps seemingly in the right direction.

US enthusiasm for CESDP has depended consistently, however, upon Washington's perception of its ability to ensure that EU initiatives constitute no threat to NATO. To this end it has repeatedly emphasised the reality of American power through NATO, the indispensability of NATO to European security, and the relative poverty of European capabilities. Some have argued that Clinton's delayed military involvement in Bosnia and his administration's undercutting of European initiatives – such as the Vance–Owen peace plan and a UN arms embargo – were 'as if there was a need to demonstrate that the Europeans could not succeed without American help . . .'.[7] That message was underscored by the profound asymmetry between European and American contributions to NATO responses to the

Sarajevo massacre in 1995 and to the Kosovo crisis in 1999. Of the 3,515 sorties flown during Operation Deliberate Force, US aircraft accounted for 65.9 per cent; four years later in Operation Allied Force that percentage increased to almost 79 per cent of the 38,004 sorties flown.

The US has also consistently sought maximum influence over the structure, objectives and missions of first a European pillar within NATO and subsequently the CESDP. In 1999 NATO endorsed in its Alliance Strategic Concept the Clinton administration's Combined Joint Task Force (CJTF) initiative as serving to reinforce the transatlantic relationship. With its central premise of separable but not separate forces, though, CJTF clearly sought to head-off an autonomous European military capability and allow the US to burden-share without surrendering leadership – not least because Europe would effectively have to borrow key US assets in a military operation. The same theme runs through the current Berlin Plus arrangements of December 2002, which set out provisions for assured EU access to NATO planning facilities for the conduct of crisis management operations. Colin Powell outlined in December 2003 exactly what, in the American view, the Berlin Plus road meant:

> When a mission comes along, if NATO for one reason or another is not prepared to accept that mission, then the EU should consider it first drawing on NATO assets. But if that also is not appropriate and the mission is within the capacity of the EU to handle alone without drawing on NATO assets, then the EU should certainly take a look at that.[8]

The US, Britain and CESDP

EU development of CESDP undoubtedly brings benefits to Washington and to NATO. In principle it encourages more effective European military spending and greater combined capabilities. Also, EU operations have already demonstrated how the Union can complement NATO, free-up US troops from peacekeeping duties and avoid US commitments in areas of little geo-strategic interest to Washington. Operation Concordia, the EU's first military peace support operation, followed on from NATO's Operation Essential Harvest and Task Force Fox in support of the Ohrid Framework Agreement in the Former Yugoslav Republic of Macedonia. Operation Althea saw the EU takeover in December 2004 from SFOR in Bosnia, and Artemis demonstrated the Union's potential usefulness in areas further afield and where the US is little interested – France serving as framework nation for intervention in the Democratic Republic of Congo in 2003.

The US exerts considerable direct influence on the balance between CESDP complementing or competing with NATO, especially through its unquestioned military supremacy and currently indispensable roles within

NATO and European security. Washington is nevertheless well aware that other countries without and within the EU would like CESDP to develop in ways that reduce American and NATO influence in Europe. US–Russia relations have periodically been sufficiently good to enable extensive mutual cuts in nuclear arsenals and more recently anti-terrorism co-operation. Nevertheless, Russia was hostile to NATO enlargement in the 1990s and its military especially is deeply concerned about the 2004 enlargement, in particular the potential for Baltic states to host not only US troops but also Ballistic Missile Defence infrastructure. While seeking a preferential relationship with NATO to influence it from the inside, Russia would still like to see its power reduced, something that was explicit in Russian initial interest in developing close cooperation with CESDP. Russia's Medium-term Strategy for Development of Relations between the Russian Federation and the European Union expressed the hope that the evolution of an ESDI might 'counterbalance, inter alia, the NATO-centrism in Europe'.[9]

Russian influence in the CESDP debate may be weak but Russia shares some common interests with the country of prime US concern – France. President Chirac's repeated references to a multipolar world mirror those of Russian President Putin and in recent years there has been evidence of intensified Franco-German diplomacy to court Russia as a potential balancing force in Europe. French officials have also spoken repeatedly of the need for autonomous European capabilities. For instance, at the same time that Chirac accepted NATO safeguards in the Nice Treaty he declared in December 2000 that 'European defence must of course be co-ordinated with the alliance, but it must, as regards its preparation and implementation, be independent.'[10] US–French relations reached a new nadir in 2003 when Paris led opposition in the UN Security Council to American interventionism in Iraq and initially blocked, together with Germany and Belgium, NATO military assistance being provided to Turkey in readiness for the conflict. All of this hardens American suspicion of CESDP as a French-led potential device to balance US power beyond the realms of trade, industry and finance.

These considerations emphasise the importance to Washington of Atlanticist allies within the EU, especially Britain. This owes in part to disappointment with Germany. Its low defence spending, difficulty in overcoming its culture of historical memory and its commitment to deeper integration have all meant that, especially since 1998, Germany has not been the partner in leadership or NATO-centric counterweight to France within the EU that was initially hoped. Moreover, Chancellor Schröder's early veto of German military involvement in Iraq seemingly indicated that Berlin was no longer the reliable ally that it had once been either. More important to US calculations, though, are that Britain is one of the two key drivers of CESDP (France being the other) and that it has a strong Atlanticist security posture, continues to stress the Anglo-American 'special relationship' and pursues an activist foreign policy that emphasises 'punching above its weight' in international affairs. All of these features potentially coincide with US interests for they give Washington particular leverage over London, and British priorities incline it to favour an outward-looking and more capable EU that operates in conjunction with NATO. For instance, British defence planning assumes that full spectrum capabilities are not required for large-scale operations because these could only conceivably be undertaken alongside the US, either as a NATO operation or as a US-led coalition. Furthermore, Washington knows that transatlantic co-operation is the *sine qua non* of Britain's balancing potentially irreconcilable tensions between its American and EU relationships. And this is something to which, as Blair made clear in his November 2004 Lord Mayor's speech, Britain remains committed irrespective of the inherent difficulties: 'Call it a bridge, a two lane motorway, a pivot or call it a damn high wire, which is how it often feels.'[11]

What roles in the development of CESDP can Britain play on behalf of Washington? There are at least three. First, Britain is a tried and trusted Atlanticist that can be relied upon to resist efforts to decouple CESDP and NATO. Britain's loyalty to the Anglo-American relationship has been demonstrated repeatedly throughout the post-Cold War era, including Major's support for the first Gulf War and Blair's commitment to military action against Iraq in 1998 and again in March 2003. In respect of an ESDI specifically, party politics and sensitivity to American opinion ensured that the Major governments were reliable bulwarks against the expansion of integration into military matters, despite potential provision for this in Article J.4 of the TEU. Blair's conversion to accepting CESDP as being conceptually compatible with NATO surprised some in Washington and drew strong criticism from irredeemable Atlanticist Margaret Thatcher, who condemned the shift as 'an act of monumental folly' taken 'to satisfy political vanity'.[12] Nevertheless, the rationale can be seen as advancing US interests for if Britain continued to adopt a blocking position, then the EU could not become the effective partner that Washington demands and a core group of countries might eventually push ahead with deeper defence co-operation regardless. Blair specifically made this argument in November 2003:

> British participation on the right terms, will ensure that European defence does indeed develop in a way fully consistent with NATO. . . . By contrast, the absence of Britain would not mean European defence didn't happen. Just that it happened and developed without us. That is not sensible for Britain, for Europe or for that matter, for America.[13]

The 'right terms', of course, are a euphemism for complementarity between NATO and CESDP, which has become a familiar British mantra. This is evident in the uncompromising language used by the Labour government on its 2005 EU presidency website: 'NATO remains the cornerstone of UK's security policy, and the only organisation for collective defence in Europe'; 'ESDP allows us to make a better and more coherent contribution to NATO'; 'ESDP will complement and not compete with NATO.'[14]

Britain's second important role in US strategy towards CESDP is, as Europe's arguably leading military power, to drive European capability improvements in ways that contribute to complement NATO. Herein Britain is a willing accomplice because it reinforces British usefulness to America and provides opportunity to secure a leadership role in CESDP through which to break the traditional Franco-German axis in favour of trilateralism. Hence Blair duly told the House of Commons in December 1999 that increased European capabilities must remain 'entirely knitted together with America on key NATO issues', and the Foreign and Commonwealth Office (FCO) embraced in its December 2003 *International Priorities* the need to strengthen the EU's capacity to undertake military operations and to 'ensure that this capacity reinforces NATO'.[15] Britain has thus emphasised the importance of the high-level EU–NATO Capabilities Group in guaranteeing that the two organisations' processes are coherent and mutually reinforcing, and the FCO has eagerly assured that 'ECAP targets are consistent with the eight top level goals identified for NATO's Prague Capabilities Commitment'.[16]

In practice Britain has assisted the US to reform NATO and has driven capability improvement within the EU, not least because it expects to pick up more expeditionary 'first-in' roles and therefore to play a lesser part in enduring operations 'where many other countries can contribute'.[17] At the 2002 Prague summit it supported NATO escaping for good the shackles of the out-of-area debate through creating the NATO Response Force (NRF). The Ministry of Defence has also supported the Bush administration in acknowledging that coalitions of the willing 'remain appropriate in many scenarios' and in calling for NATO to 'continue to become more flexible – institutionally, politically and militarily – if it is to deliver effective force relevant to the evolving security environment'.[18] Meantime, following Anglo-French sponsorship at the Le Touquet summit in February 2003 of a new European Defence Agency (EDA), Britain pushed hard against countries such as Germany and Italy for national Defence Ministers to sit on the EDA's Steering Board as a means of ensuring greater political commitment to effective outcomes.[19] Britain called also for greater liberalisation of the EU defence market and proposed in February 2004, in conjunction with France and Germany, the creation of EU rapid reaction units comprised of joint battlegroups. This was characteristically accompanied by British assurance that the battlegroup initiative 'does not compete with the NATO Response Force but is designed to be complementary and mutually reinforcing, with each providing a positive impetus for military capability improvement'.[20]

Britain also has a responsibility within US policy to steer the CESDP political process in directions that overcome traditional European introspection and to develop EU strategy in ways that maximise the Union's potential contribution to US global policies. There is again some evidence of success in this respect. Consider the EU's first ever European Security Strategy (ESS), adopted in December 2003. The document was heavily influenced by Blair's former foreign policy adviser Robert Cooper and bore unmistakable British fingerprints. The ESS included association with the US doctrine of pre-emptive defence, the need to develop a capability for robust intervention, and the importance of combating threats from far beyond the EU's near abroad, including Weapons of Mass Destruction (WMD) proliferation. This tellingly echoed the FCO's *UK International Priorities*, released that same month, which emphasised that 'the focus of our security and defence policy will be on understanding and countering new threats, often from non-state actors empowered by new technologies, and originating outside Europe'.[21]

Britain is in many respects a natural pole for other EU states with Atlanticist security postures to gather around. It is certainly the only member of the 'Big Three' that currently can convincingly assuage concerns that CESDP will neither weaken NATO nor encourage American military disengagement from Europe. Also, the George W. Bush administration is well aware that the stronger this pro-NATO caucus is within the EU then the more likely it is that Anglo-American pressure will succeed in keeping CESDP in a form acceptable to the US. It has thus moved to strengthen ties with key Atlanticist 2004 EU entrants such as Poland. The Bush administration's crude disaggregation policy during the Iraq intervention controversy was disliked by most EU member states and prospective members. Nevertheless, dividing the EU undermined Franco-German objectives for deeper European defence integration, and forcing states into a 'with us or against us' stance underlined the importance and strength of the 'America factor' in EU security aspirations. As Estonian Prime Minister Siim Kallas observed,

> I do not want war [in Iraq]. . . . But I believe we must pick a side. And I believe it is the side where the United States is . . . isn't it absurd to imagine that hiding behind the wardrobe and wagging a reproaching finger at America would better ensure our defence from Stalin Jr. than being an outright public friend and ally of the United States?[22]

Furthermore, some Bush administration policies constitute something of a 'pre-emptive strike' against the Europeanisation of 2004 EU entrant states' foreign and security policies. PfP and more especially the macro NATO enlargement of 2004 tied many of the EU's new member states into NATO and its military programmes before they became full members of the

Common Foreign and Security Policy (CFSP) and CESDP. In addition, the Bush administration plans a reduction of up to one-third of its troops based in Europe over the next seven to ten years as part of its global military repositioning exercise and ongoing Base Closure and Realignment rounds. Within this it has indicated that it will move some military assets from Germany eastwards into Poland and even into Baltic NATO members, thereby sending a strong message of US support for these countries and encouraging their Atlanticist disposition within the EU.

US policy – how sustainable?

The US has to date largely succeeded in ensuring that CESDP develops complementarily with NATO. It has been ably assisted by Britain in this task. Blair pre-empted CESDP spill-over into collective defence by first ensuring that the EU did not absorb the WEU's mutual defence commitment in the Nice Treaty, and later forced a compromise in December 2003 over two key defence-related provisions in the draft EU constitution. The section on mutual military assistance was revised to acknowledge NATO's continued responsibility for collective defence, and structured co-operation on defence was limited to a European defence avant-garde group increasing military capabilities. The US and Britain have also largely worked together to push successfully for Turkey to begin pre-accession negotiations for EU membership. This provides an obvious further bulwark against a French-led group of countries being able to use CESDP to weaken NATO.

There seems little immediate risk of CESDP being decoupled from NATO, especially given that European ability to meet anything more than light-end Petersberg Tasks remains seriously compromised by key military deficiencies, including special operations forces, transport aircraft and sophisticated C_4I systems. Potential dangers from Washington's perspective arising from a competitive CESDP lie in the medium to long term. The most obvious comes from a 'core Europe' resolved upon autonomous capabilities and deepening defence integration, possibly to include mutual defence. The draft EU Constitutional Treaty demonstrated that this is a real possibility, as did the so-called Chocolate Summit in April 2003. Here Belgium, France, Germany and Luxembourg announced agreement to deepen defence collaboration and proposed the creation within one year of a permanent EU operational military planning cell (as an alternative to NATO) in Tervuren. Britain subsequently contained this initiative by securing an effective national veto over military operations and getting the December 2003 European Council to agree that SHAPE and national headquarters remained the main options for conducting EU-led operations. Still, though, US Ambassador to NATO Nicholas Burns warned that developments within CESDP represented 'one of the greatest dangers to the transatlantic relationship'.[23] This is the 'thin end of the wedge' argument whereby Europe secures greater autonomy of the US through gradual erosion of NATO primacy. And it is not without some justification. For instance, shortly after the EU's first use of SHAPE planning and command capabilities in Operation Concordia some countries, notably France, were reportedly unhappy that the EU command element in AFSOUTH was not fully under EU control and that many of SACEUR's functions were performed by AFSOUTH.

A more subtle challenge lies in threats to interoperability posed by America's technological supremacy and in potential consequences of Europe's drive for improved procurement and R&D collaboration. On the one hand, the ability of NATO Europe and the EU to add military value to unilateral US assets is already very limited. This was clearly demonstrated in the Bush administration's decision to largely ignore NATO in its Afghan and Iraqi interventions and in Rumsfeld's politically misjudged assertion of the dispensability of Britain's contribution in Iraq. Neither is this situation likely to change. Pressing ahead with Joint Vision 2020 and absorbing spiralling costs in Iraq mean that US defence expenditure is likely to equal that of the rest of the world combined by the end of 2006. On the other hand, well-intentioned initiatives such as the EDA risk intra-European defence industry consolidation and Eurocentric procurement programmes rather than liberalisation of the transatlantic defence market. This would make interoperability with US forces more difficult and would exacerbate the currently difficult balancing act in military procurement performed by some of the EU's Atlanticist-orientated countries. Britain is a prime example of the latter. It is the only level 1 partner in the American Joint Strike Fighter project and has politically offset this by its commitment to the European A400M project for heavy airlift provision. Moreover, Britain is also unlikely to want to surrender its privileged defence relationship with the US. While this may give American administrations leverage, it could also undermine EU countries' collaborative efforts to provide the military capabilities that the US demands for reasons of burden sharing.

The problem of balancing between European and US relationships also raises questions about the ability of America's foremost European ally, Britain, to secure and maintain leadership of the CESDP process. Blair, of course, continues to maintain that Britain should no longer be 'mesmerised by the choice between the US and Europe. It is a false choice.'[24] Critics from all sides are ever less convinced of this. They variously question the sustainability of Blair's 'transatlantic bridge', the reality of an Anglo-American 'special relationship' and London's ability to avoid choosing without compromising influence with either party in an era of growing transatlantic differences. As Hugo Young put it, 'instead of being Europe's voice in America and America's voice in Europe, Britain runs the risk some day soon of having a small voice and smaller audience in either place'.[25] Certainly Britain's support of the US in Iraq temporarily dislocated the Anglo-French St Malo *rapprochement* and undermined London's drive for trilateralism within the

EU by pitting it against Berlin and Paris. Also, Britain's 2005 EU presidency made little progress in unblocking the budget process and resolving the crisis over the popularly rejected Constitutional Treaty. Rather, London is again confronting the Franco-German axis on wide-ranging contentious issues, including CAP reform, Britain's budget rebate and the future of the European social model. Furthermore, even if Britain could secure greater trilateralism within CESDP it would still have to deal with the sensitivities of other EU states. Evidence of greater British–French–German foreign and security policy co-ordination before the Iraq war was greeted with alarm and anger in some other European capitals. Spanish Foreign Minister Ana Palacio proclaimed that 'nobody should be allowed to kidnap the general interests of Europe', and Spain, Italy, Belgium, the Netherlands and High Representative for CFSP Solana all insisted on being invited in November 2001 to a Downing Street dinner originally intended by Blair for his French and German counterparts.[26]

It is a bitter irony, however, that the more serious threats to American interests in CESDP–NATO complementarity derive currently from US policies. For a start, there is a tension in American objectives between ensuring NATO primacy and US European influence and pushing the Europeans to develop improved military capabilities. Put simply, the more capable the Europeans become, the more resistant to Washington's preferences they might be. This has happened already in the transatlantic economic relationship and even Britain has ambitions for CESDP beyond being a junior partner of NATO and a facilitator of transatlantic co-operation. As Blair told the European Parliament on 23 June 2005: 'a defence policy is a necessary part of an effective policy. . . . A strong Europe would be an active player in foreign policy, a good partner of course to the US but also capable of demonstrating its own capacity to shape and move the world forward.'[27] The George W. Bush administration has repeatedly demonstrated the dilemmas posed by this long-standing tension in US policy and even a current inclination to favour the EU failing to be an effective burden-sharer over its being a 'too capable' partner. For instance, it initially opposed the EU's development of the Galileo satellite navigation system and responded to the Helsinki Headline Goals and development of the EURRF by launching parallel processes within NATO. NATO emphasises that these initiatives are complementary to the EU's but the NRF undoubtedly nullified the EURRF's initial range advantage over A5-bound NATO, and the Prague Capability Commitments are liable to ensure that the NRF will 'glean the lion's share of Europe's limited resources'.[28] The Bush administration's Iraq policy also knowingly rendered the EU ineffective by dividing its member states, the Union's CFSP being conspicuous by its absence.

There is a second inconsistency within US demands, this time between wanting a more capable Europe and warning of a *de facto* technological decoupling of the Atlantic Alliance and loss of interoperability. These dangers could be partially mitigated by greater technology transfer, more co-ordinated procurement and reduced protectionism of defence markets – something that US administrations have recognised. On 24 May 2000 Madeleine Albright announced the Defence Trade Security Initiative aimed at improving US technology and arms transfers to America's closest allies. Yet at the same time the Clinton administration subsidised the US defence industry's post-Cold War reconfiguration by $16.5 billion, including $1.3 billion within the Technology Reinvestment Programme designed to promote dual-use technologies. Also, for all its criticism of potential Euro-centric procurement programmes, the US is equally guilty of favouring its national defence industry. The US spends currently under 2 per cent of its defence budget beyond its borders – of which half is concentrated in the United Kingdom. Congress poses a further challenge to transatlantic defence industry collaboration. It guards a formidable set of protective measures against FDI, joint ventures and exports to third parties, including the Buy American Act, the Arms Export Control Act and the International Traffic in Arms Regulations. It is also highly sceptical about transferring technology and sharing intelligence with European partners. For instance, Republican Senator Jon Kyl responded to NATO Secretary General George Robertson's plea in early 2002 for technology transfer to improve European capabilities and interoperability by reproaching the Europeans for developing Airbus to compete with Boeing's air transports and asserting that sharing technology with the Europeans could lead to it leaking to rogue states.

Perhaps the biggest current danger, though, at least in terms of a CESDP that competes with NATO, is of US policies encouraging reactions within the EU that promote the attractions of a pro-autonomy core Europe and weaken the hand of Atlanticists. Preconditions for this are established. The EU is ambitious of a strengthened role in international security and has well-documented differences with the US over preferred security tools, modes of operation and willingness to abide by multilateral agreements. EU resentment about US failure to consult consistently in developing policy is likewise clear. European Commission Vice-President Leon Brittan warned explicitly in September 1998, 'that co-operation with the European Union does not mean simply signing up the European Union to endorse, execute, and sometimes finance, United States foreign policy'.[29] In this context repeated disaggregation policies of 'Old' and 'New' Europe risk a French-led grouping of EU states being able to use American interference and 'cherry picking' as a rallying point for a more tightly integrated EU. Similarly, recent US sidelining of NATO and its relegation to being a repository of coalitions of the willing could scarcely be better calculated to feed a European search for security alternatives. The US exacerbated this in France's case by using the latter's absence from NATO military command to overcome French-led opposition to deploying NATO assets to Turkey prior to intervention in Iraq. More significant still is that US policy has seemingly been driving traditionally pro-NATO Germany further towards supporting a more autonomous European capability. Chancellor Schröder indicated this with surprising bluntness in February 2005 in his statement that NATO was 'no longer the primary venue where transatlantic partners discuss and coordinate strategies'.[30]

The Bush administration is alive to some of these dangers and has also been somewhat chastened by its ongoing experience in Iraq, at least insofar as needing assistance in pacification and reconstruction. The days that saw Condoleezza Rice recommend that the US 'punish' France and 'forget' Germany are gone, if not forgotten. Bush quickly indicated a change of tone for his second term. In his inauguration speech he stated of US allies: 'we honor your friendship, we rely on your counsel, and we depend on your help. . . . The concerted effort of free nations to promote democracy is a prelude to our enemies' defeat.'[31] He also pointedly made NATO headquarters his first stop on his 'listening tour' to Europe in February 2005 and directly contradicted Schröder by emphasising that NATO 'is the vital relationship for the United States when it comes to security' and that it is 'a place for us to have meaningful dialogue'.[32]

Beyond renewed stress on transatlantic co-operation and shared values, the Bush administration has also retreated from policies of crude disaggregation in Europe in favour of fence-building, irrespective of continued transatlantic differences over a host of multilateral agreements and specific issues such as policy towards Iraq and EU lifting of its arms embargo on China. This has reduced the polarisation of strategic visions of CESDP and better enabled a return to the studied ambiguity that previously allowed all the major players to endorse cautiously incremental capability building, the US included. Fence-building has also weakened the arguments of states favouring a core Europe in defence issues and ameliorated, however temporarily, the tensions faced by a number of EU states between Atlanticist-orientated defence postures and EU-centred economic priorities.

Finally, the Bush administration has been favoured by recent political events in Europe. Its staunchest European ally, Blair, secured a third term in May 2005. Schröder was eventually forced from the German chancellorship following German elections in September 2005. And Chirac has been seriously wounded by the French popular rejection of the EU Constitutional Treaty and faces an uphill struggle ahead of French elections in 2007. This undermining of momentum behind a Franco-German axis that was increasingly critical of NATO and supportive of deeper EU defence integration comes on top of poor diplomacy during the Iraq controversy that antagonised a number of CEECs. Even more propitious for the Bush administration is that the new German Chancellor, Angela Merkel, is more pro-Atlanticist than Schröder. Although limited by the constraints of a Grand Coalition government and unenthusiastic about Turkish EU membership, Merkel is at least likely to quieten German criticism of NATO, and be less beholden to French President Chirac and more amenable to British preferences in CESDP.

Conclusion

The Pentagon especially might suffer Francophobia and successive administrations have shown themselves to be nervous of a creeping advance of an autonomous ESDI. Yet the Berlin Plus arrangements have provided opportunity to embed NATO–CESDP co-operation within accepted parameters and to set a benchmark for US policymakers and EU Atlanticists to defend. This defence will be vigorous when needed and herein Britain is likely to remain America's loyal lieutenant, not least because its wider foreign policy rests precariously on transatlantic co-operation being sufficiently good to avoid, or at least defer, painful choices between its US and European relationships. Meantime there is little prospect of the radical improvement in European military capabilities sufficient to operationalise 'autonomy' in anything other than light-end Petersberg Tasks. Moreover, the EU is likely to be preoccupied for some time with internal problems of the budget, a substantially larger membership and the rejected Constitutional Treaty. What effort is dedicated to CESDP will therefore probably inch it in directions acceptable to America, especially given the 2005 election results in Britain and Germany and Chirac's weakened political position.

American fears of an autonomous Europe undermining NATO through a strong and competitive CESDP are therefore at least premature. To discern any such threat requires uncertain extrapolation from recent trends and the projection of these into the medium- to long-term future. For example, it is likely that current tensions between US demands for NATO primacy, burden sharing and preserving US influence in Europe will become more pronounced if the EU continues developing CESDP capabilities slowly. The overall tenor of transatlantic relations will have an important bearing too. The Bush administration needs to invest more than words in NATO to demonstrate its commitment to the organisation and the organisation's relevance to post-9/11 US policy and international security. The Bush administration likewise needs to avoid a repeat of the grave transatlantic tensions caused by its unilateral decision-making over Iraq. Should it fail in these tasks, then it will weaken the position of EU Atlanticists and potentially provide impetus to that which it most fears: a core Europe in defence matters dedicated to autonomous capabilities.

Still, though a meaningful 'core Europe' in defence matters that insists on autonomy is theoretically conceivable, it remains difficult to envisage one developing in the foreseeable future, especially without a fundamental shift in British policy. The most likely immediate scenario, and one that does pose danger to Washington's interests and to transatlantic relations, is

CESDP remaining too weak. US criticisms about loss of interoperability, limited European R&D and weak capability return on European military spending risk becoming a self-fulfilling prophecy unless the US becomes more willing to share technology and invests considerable effort in liberalising the transatlantic defence market. The counterpart to that equation is even more important and less promising: to date the majority of EU states have failed to translate commitments into military assets. Nor are they likely to do so to a meaningful extent in the foreseeable future given that most of them labour variously under Stability and Growth criteria, costs of the European social model, continuing adaptation to post-communist economies, lack of international competitiveness and continued attachment to sovereignty in defence matters. The likely product of this is threefold. First, renewed American criticism of European inability/unwillingness to fulfil security strategies and commitments pledged on paper. Second, continued transatlantic drift as Europe fails, ironically assisted by American policies, to provide the US with good reason or opportunity for meaningful military co-operation. Third, the Bush administration being cast back upon its natural preference for coalitions of the willing and privileged partners, which in turn weakens NATO and CESDP and feeds the transatlantic cycle of tension and recrimination. This cycle is breakable with concerted effort and compromise but at present it is unclear whether the US or the EU have the political will to do so.

Notes

1 Robert Gates, *From the Shadows* (New York: Simon and Schuster, 1996), p. 552.
2 James Baker, cited in M. Walker, 'Variable Geography: America's Mental Maps of a Greater Europe', *International Affairs* 76:3, 2000, p. 460.
3 Declaration on US–EC Relations, htttp://www.useu.be/Transatlantic/transdec.html
4 Ambassador Eizenstat, cited in T. Frellesen, 'Processes and Procedures in EU–US Foreign Policy Co-operation: From the Transatlantic Declaration to the New Transatlantic Agenda', in E. Philippart and P. Winand (eds) *Ever Closer Partnership. Policy-making in EU–US Relations* (New York and Oxford: Peter Lang, 2001), p. 333.
5 R. Seitz, *Over Here* (London: Weidenfeld & Nicolson, 1998), p. 322.
6 M. Albright, 'The Right Balance Will Secure NATO's Future', *Financial Times*, 7 December 1999; remarks by US Secretary of Defence at Munich Conference on European Security Policy, http://www.defenselink.mil/speeches/2001/s20010203–secdef.html; comments by US Secretary of State Colin Powell, 2 December 2003, http://www.useu.be/Categories/Defense/Dec0203PowellOSCE.html
7 S. Hoffmann, 'The United States and Europe', in R.J. Lieber (ed.) *Eagle Adrift: American Foreign Policy at the End of the Century* (New York: Longman, 1997), p. 190.
8 Comments by US Secretary of State Colin Powell, 2 December 2003, http://www.useu.be/Categories/Defense/Dec0203PowellOSCE.html

9 Medium-term Strategy for Development of Relations between the Russian Federation and the European Union (2000–2010), 1999, unofficial translation, http://europa.eu.int/comm/external_relations/russia/russian_medium_term_strategy/
10 'Chirac springs surprise on defence', Thursday, 7 December 2000, http://news.bbc.co.uk/1/hi/world/europe/1060172.stm
11 Tony Blair, speech at the Lord Mayor's Banquet, 15 November 2004.
12 Margaret Thatcher, cited by R. Harris, 'Blair's "Ethical" Policy', *The National Interest* 63, 2001, p. 34.
13 Blair, speech at the Lord Mayor's Banquet.
14 'European Security and Defence Policy', http://www.eu2005.gov.uk/servlet/Front?pagename = OpenMarket/Xcelerate/ShowPage&c = Page&cid = 1107293537693
15 House of Commons debate, *Hansard*, 13 December 1999, col. 39. Command 6052, *UK International Priorities*, p. 35.
16 Foreign and Commonwealth Office, 'Capabilities for ESDP', http://www.fco.gov.uk/servlet/Front?pagename = OpenMarket/Xcelerate/ShowPage
17 Command 6041-II, 'Delivering Security in a Changing World: Supporting Essays', December 2003, p. 3.
18 Ibid., p. 1.
19 D. Keohane, 'Europe's New Defence Agency', policy brief, June 2004, http://www.cer.org.uk, p. 5.
20 Foreign and Commonwealth Office, 'Capabilities for ESDP', http://www.fco.gov.uk/servlet/Front?pagename = OpenMarket/Xcelerate/ShowPage
21 Command 6052, *UK International Priorities. A Strategy for the Foreign Office*, December 2003, p. 13.
22 Siim Kallas, cited by T. Valasek, 'The "Easternization" of ESDP after EU Enlargement', *New Defence Agenda*, Winter 2004, p. 53.
23 Cited in 'Defensive Reactions', *Guardian*, 18 October 2003.
24 Speech by Blair at the Lord Mayor's Banquet, 22 November 1999, www.fco.gov
25 Hugo Young, 'Comment and Analysis: Perhaps a Russian-British lobby against war on Iraq? If Blair persists in speaking for Bush, his voice will get smaller and smaller', *Guardian*, 19 February 2002.
26 Palacio, cited in K. Larres, 'Time for a threesome', *The World Today*, 60: 4, 2004, p. 8.
27 Speech by Prime Minister Blair to the European Parliament, 23 June 2005.
28 E.G. Book, 'Creating a Transatlantic Army: Does the NATO Response Force Subvert the European Union?' 29 November 2003, *Sicherheit und Verteidigung*, http://www2.dias-online.org/direktorien/sec_def/031129_21
29 L. Brittan, 'Europe and the United States: New Challenges, New Opportunities', speech to the Foreign Policy association, New York, September 1998, http://www.eurunion.org/news/speeches/1998/1998index.htm
30 'Germany Urges NATO Reform and Rethink of Transatlantic Ties', *Agence France Presse*, 13 February 2005.
31 President Bush, inauguration speech, 20 January 2005.
32 'President and Secretary General de Hoop Scheffer discuss NATO meeting', press conference, US State Department transcript (22 February 2005), http://www.state.gov/p/eur/rls/rm/42523.htm

8

BEYOND NATO?

The European Security and Defence Project

Jolyon Howorth

The Alliance is dead: long live the Alliance

Since the end of the Cold War, the Atlantic Alliance has been declared dead – or on life support – more times than anybody can count. At the same time, it has proved remarkably resilient at re-inventing itself. Two recent books, both written by NATO 'insiders', encapsulate the parameters of the debate on NATO's current state of health. Rupp presents a sophisticated and compelling argument that NATO has entered a period of terminal decline.[1] Rynning presents a robust analysis of the objective and subjective reasons for continuing to believe that the Alliance has a strong future ahead of it.[2] Alliances, both history and political science suggest, rarely outlive the demise of the threat against which they were originally constructed. The biggest single challenge to NATO since 1989 has, of course, been that of finding a new security role for its *European* member states. Under US nuclear leadership during the Cold War, the Europeans were assigned specific tasks which meshed tightly with the Alliance's defensive grand strategy. The military disparities between the US and the Europeans which Robert Kagan polemically deplored in terms of 'power and weakness' derived largely from geo-strategic considerations. In order to carry out its Cold War role as the protector of Western Europe, the US was obliged to prioritise *power projection* capabilities. Since it simultaneously aimed to stabilise much of Asia and the Middle East, this capability became truly global. The Europeans, on the other hand, were confined to static line defence within their own limited geo-strategic area. With the Cold War confined to history, the real question implicit in Kagan's taunts of 'weakness' was whether the Europeans could – or indeed would wish to – develop power projection capabilities of their own. And to what end? These questions were initially seen as being intimately linked to the process of NATO transformation.

NATO was founded, in the words of the organisation's first Secretary General, Lord Ismay, 'to keep the Americans in, the Russians out and the

Germans down'. After 1989, all three elements lost salience. America's strategic interest in Europe declined steadily and, to a large extent, the (reluctant) US engagement in the Balkans was driven as much by concerns about NATO's own credibility as by any strategic imperative. The 2004 *US Global Posture Review* represents the effective end of nothing more than a symbolic American military presence in Europe.[3] The Russians, far from being kept 'out' – or even at arm's length – have been invited 'in', first via the Partnership for Peace (PfP) programme and later via the NATO–Russia Council. As for Germany, the drive to keep it 'down' is more a function of that country's own historico-cultural *angst* than of any desire on the part of its allies to deny its natural place at the heart of Europe. History, in 1989, went into fast-forward mode. NATO tried valiantly to re-invent itself and to evolve from a collective defence to a collective security organisation. The focus on the former massive identifiable *threat* from the USSR was replaced by a concentration on multifaceted *risks* posed by new vectors of instability – regional crises, ethnic tensions, terrorism, WMD proliferation. But at the heart of NATO's dilemmas remained the question which had given rise to the organisation in the first place: who was to be responsible for the delivery of European security?

From a European Security Identity to a European Security and Defence Policy

Long-standing quarrels between the European allies and the US over 'burden-sharing' were notionally resolved when the former, in the mid-1990s, agreed to take on more responsibility for regional security through the creation of the European Security and Defence Identity (ESDI). This was an arrangement whereby European-only forces, under a European command chain, could be 'identified' from inside the Alliance, equipped with borrowed American assets and entrusted with crisis management missions in Europe's backyard, thereby relieving US forces for more urgent tasks elsewhere around the globe . The arrangements hinged on the 'Berlin Plus' procedures allowing the EU to enjoy 'assured access to NATO planning', 'presumed access to NATO assets and capabilities' and a pre-designated Europeans-only chain of command. All this was agreed in principle at a NATO summit in Berlin in June 1996. The devil proved to be in the detail and it took six years of hard bargaining (the 'Berlin Plus negotiations') to nail down those details. This was finally achieved in December 2002 when NATO and the EU signed a joint declaration announcing a 'strategic partnership [. . .] founded on our shared values, the indivisibility of our security and our determination to tackle the challenges of the new Century'.[4] Few, however, were under any illusions that the principles of shared risks and shared responsibilities on which NATO had stood strong for forty years still held. The US armed forces were qualitatively in a different league from those of the Europeans, as was conclusively demonstrated in the Gulf War of 1991 and in the Kosovo campaign of 1999. Interoperability between US and European forces was increasingly problematic. An Alliance whose partners cannot fight effectively together is like a motor car with no gearbox. When, on the morrow of 9/11, NATO for the first time in its history invoked 'Article 5',[5] the response from the US side was 'Don't call us, we'll call you.' Many analysts concluded that this incident represented the final nail in NATO's coffin. The Europeans, drawing the logical conclusion, stepped up their efforts to create, not an identity inside NATO, but an autonomous European Security and Defence Policy (ESDP) outside it.[6]

Tony Blair's historic deal with Jacques Chirac at St Malo in December 1998 is universally recognised as a major turning point in European security policy. Its impact on EU–US relations cannot be overstated. St Malo represented a threefold challenge to all previous assumptions about the security relations between the two sides of the Atlantic. First, by using the dramatic word 'autonomous', it launched a vigorous debate (still ongoing) about the future coherence and unity of the Atlantic Alliance. Second, by conferring upon the EU a political role in strategic decision-making and, eventually, in matters of war and peace, it changed in radical ways the nature of the EU both as an international actor and as a (hitherto) exclusively civilian actor. Finally, by associating the UK so tightly with a project which appeared to vindicate five decades of French agitation in favour of greater strategic independence for Europe, it not only raised more than the odd eyebrow in Washington but also predicated the entire future of the ESDP – as it was soon to become known – on the maintenance of a good working relationship between Paris and London. When that relationship has flourished (roughly between St Malo and summer 2001 and again since summer 2003), ESDP has made progress. When the relationship faltered (2001 to 2003), ESDP has faltered with it.

St Malo raised a range of major problems which the EU collectively and the member states individually have been grappling with ever since. Most of them were sparked by the implications of the word 'autonomous'. The institutional implications were rapidly resolved and within eighteen months the EU had successfully implanted in Brussels a raft of new bodies: the High Representative for the CFSP (HR-CFSP – Javier Solana) and his advisory Policy Unit (PU); the Political and Security Committee (COPS, from the French acronym) comprising ambassadors from each member state; the European Union Military Committee (EUMC) formally made up of the Chiefs of the Defence Staff of the member states; and the EU Military Staff (EUMS) comprising some one hundred and fifty senior officers from across the Union. The emergence of the EU as a security actor immediately raised the question of the relationship between these new political-military institutions and their equivalents in NATO. At the Nice European Council meeting of December 2000, pragmatic arrangements were institutionalised whereby regular meetings of the Political and Security Committee (PSC) and the North Atlantic Council (NAC) would be complemented by attendance at the meetings of the one by the Chairperson of the other and by semi-permanent dialogue between both the Military Committee and the Military Staff of both organisations. These meetings have been held frequently ever since, but participants agree that the atmosphere remains strained as a result of the asymmetry between the respective profiles, culture and expertise of the two sides.

More problematic was the resolution of the EU's working relationship with NATO. This involved two interlocking issues. The first was the implementation of the 'Berlin Plus' arrangements for transfer to and from the EU of NATO military assets. The second was the involvement in ESDP of non-EU NATO members such as Turkey and Norway. The latter problem dominated the headlines, while the former was tackled from behind closed doors. Turkey was particularly disturbed by the ESDP project for three reasons. First, while as an associate member of the WEU from 1992 onwards Turkey had been fully involved in intra-European security discussions, ESDP offered no such facility. The six non-EU European member states of NATO[7] not only found themselves excluded from the EU's COPS, but watched from the sidelines as four formerly neutral countries (Austria, Finland, Ireland and Sweden) took up full seats on that committee. Second, Turkey feared that it was witnessing a process whereby the US (in which Ankara had enormous faith) progressively transferred responsibility for European security to the EU (in which Ankara had very little faith). Third, this was all the more unacceptable for the Turks in that most scenarios for armed conflict and crisis management in the European theatre seemed to be situated in the south-eastern parts of the continent, an area which Turkey regarded as its own 'backyard'. In particular, Turkey feared the use of ESDP military assets to intervene in Cyprus on behalf of an EU member state, Greece. Ankara therefore decided, in spring 2000, to block the Berlin Plus process by threatening to veto the transfer to the EU of those indispensable NATO assets without which the EU could hardly embark on military operations at all. The challenge of securing the EU's own 'near-abroad' involved not only a new relationship with NATO and with the US, not only a radical re-configuration of available EU military assets, but also a new deal with the EU's immediate neighbours.

It took almost three years of high-level negotiations, led mainly by the UK, to reach an agreement acceptable both to Ankara and to Athens. On 16 December 2002, the EU and NATO issued a 'Declaration on ESDP', announcing the strategic partnership between them, and asserting that, while the EU would ensure 'the fullest possible involvement of non-EU European members of NATO within ESDP', NATO, for its part would guarantee the EU 'assured access to NATO's planning capabilities'. The Declaration also announced 'arrangements to ensure the coherent, transparent and mutually reinforcing development of the capability requirements common to the two organisations'.[8] On paper, at any rate, the EU had finally solved the conundrum of its complex relationship with NATO. However, the fine print of the EU–NATO relationship remains highly classified and the specific details of the Berlin Plus arrangements have never been made public. Many analysts believe that, in the event of two crises arising in the world simultaneously, one vital to the US and one vital to the EU, the likelihood of the latter being able to count on the availability of key military assets belonging to the former is slim. Such a situation stimulates the European drive for real autonomy in the area of military capacity. This remained the fundamental challenge as the EU entered the twenty-first century.

The question of European autonomy was also raised at politico-diplomatic level by controversy over NATO's so-called 'right of first refusal'. Shortly after St Malo, leading US spokespersons insisted that, as and when a given crisis scenario began to loom, NATO should first of all decide whether it wished to be involved – and *only if it did not* would the task of crisis management be handed over to the EU. This interpretation was fiercely resisted by France and one or two other countries, while being discreetly supported by the UK and a handful of allies. As the EU began, in 2002, to emerge as a potential crisis management actor, a version of 'right of first refusal' kicked in. Both with respect to the EU mission in Macedonia (Operation Concordia in April 2003 – see below) and with respect to the EU's takeover of NATO's peace-keeping mission in Bosnia (Operation Althea), a *de facto* NATO decision preceded the EU's formal assumption of responsibility. This tension persists and is likely to produce more tension within NATO as ESDP matures.

At the geo-strategic level, the main issue facing the two bodies has been that of their overall complementarity. The initial US response to the St Malo Declaration, as formulated by Secretary of State Madeleine Albright, stressed US opposition to any 'decoupling' of the two sides of the Atlantic. Albright stated that 'European decision-making [should not be] unhooked from broader Alliance decision-making'. The unfortunate symbolism of this stricture was more significant than its practical reality. To state, on behalf of a politico-military alliance, that it should not be permitted to fall apart is rather like the vicar announcing his opposition to sin. The statement had two equally unfortunate connotations. The first was the implication that the EU allies were *actively seeking* to 'unhook' the Alliance, an accusation which had little basis in fact and which came across rather like the fretting of an over-anxious parent, concerned lest the adolescent child be led astray by life's temptations. The second connotation was that the US *would not permit* decoupling, this one being even more offensive in that it appeared to involve some measure of threat. In the end, whether the Alliance remains in business or 'decouples' will be a consequence of the respective desires on both sides of the ocean to keep it in business.

That desire will be tested most strongly by the EU's eventual – and crucial – decision as to whether or not to align itself formally with what has become a new US global strategic focus. That is a debate which is currently ongoing. Its first serious test – in Iraq since 2003 – hardly augured well.

A third set of problems for EU–US relations arose at the military-operational level. The first of these had to do with another of Madeleine Albright's 'no go areas': military duplication. It was perhaps ironic that US fears, immediately after St Malo, focused on the dangers of the EU coming to *rival* the US in military hardware. This fear was not unconnected with the other ambition expressed in the St Malo Declaration: 'a strong and competitive European defence industry and technology'. At the time, European defence and aerospace industries were engaged in a major process of restructuring and cross-border mergers, leading to predictions that a single massive multinational company might emerge – a company which the media had already named the European Aerospace and Defence Company (EADC). The US need not have worried. EADC did not happen. And, within eighteen months, it was clear that, far from the EU rivalling the US in military hardware, the real problem was that Europe was losing ground fast in the race to produce a new generation of weapons systems. This process was aggravated by 9/11 which inevitably led to even greater procurement efforts on the part of the US. Yet 9/11 carried another message with respect to duplication. One of the initial arguments against duplication had been that it was unnecessary for the Europeans to procure systems which already existed within NATO (meaning in the US). These could, if necessary, be borrowed by the EU under the Berlin Plus procedures. What 9/11 made clear, however, was that US assets were henceforth likely to be in short supply. It also underlined the obvious truth that, if there was Alliance scarcity, particularly in strategic systems – long-range transport, logistics, command and control – it made good sense to fill those gaps. Since 2001, the EU has struggled to implement the European Capability Action Plan (ECAP), designed to identify the major shortcomings in Europe's military capacity. In summer 2003, the ECAP process began to focus more on qualitative approaches and criteria. Project groups were established to focus on solutions such as leasing, multinationalisation and role specialisation. The EU moved towards the recognition of 'coordination responsibility' for key procurement projects: Germany, strategic air lift; Spain, air-to-air refuelling; UK, headquarters; Netherlands, PGMs for delivery by EU F-16s. Little by little, the EU was beginning to acquire the accoutrements of serious military capacity. How did all this relate to developments within NATO itself?

Developments in NATO

In November 2002 at a summit meeting in Prague, the Alliance, triathlon-style, announced a triple leap forward. First, breaking with its long-standing self-imposed geographical limitations, NATO declared it was 'going global'.[9] Second, it announced that a new priority focus was to be the 'threat posed by terrorism'. Third, it vowed to establish a rapid reaction force, the NATO Response Force, comprising up to 20,000 troops for instant deployment anywhere in the world. The Allies, temporarily basking in the harmonious glow of UNSC Resolution 1441 which had called on Saddam Hussein to disarm, laid aside any national concerns or inhibitions and adopted the Prague Declaration. All were keen to demonstrate – to whomever might be paying attention – that they were still, collectively, very much in business. Yet by an imperceptible process of osmosis, the Alliance, which had initially been designed to deliver US commitment to European security, had thus suddenly been transformed into an Alliance essentially configured to deliver European commitment to US global strategy.

After Prague, the Alliance struggled to illustrate its new global remit. The obvious example of success was the International Security Assistance Force (ISAF) in Afghanistan. This mission, initially led by individual member states (UK, Turkey, Canada and Germany), was commanded directly by NATO after August 2003. This was the first mission outside the Euro-Atlantic area in NATO's history. On the surface, it seemed an excellent example of NATO's 'out-of-area' potential. However, on closer scrutiny, the evidence is ambiguous. First, Afghanistan is the one place on earth where nobody – except Al Qaeda and the Taleban – objected to a US military presence. Achieving consensus on a NATO Afghan mission is hardly challenging. But the ISAF mission itself was very unambitious – stabilising Kabul in order to give President Karzai a fighting chance to generate a political process, and establishing a handful of Provincial Reconstruction Teams (PRTs) to reach out into the provinces. Even then, NATO member states proved unwilling to send more than an absolute minimum of troops and resources in order to reach these limited goals. This was hardly a revolution in Alliance affairs. To stabilise Afghanistan properly could require as many as 250,000 troops. The 8,000 deployed in 2005 – drawn from forty-seven nations – could do little more than fly the flag. Looking beyond Afghanistan, analysts speculated about other global hot-spots where NATO might deploy. Kashmir? Chechnya? Gaza? Darfur? Korea? As the Cook's tour of the globe proceeded, political objections abounded. The simple truth is that, for one reason or another, a now expanded Alliance with twenty-six member states will be hard pressed to reach political consensus – let alone unanimity – on deployment to any one of the earth's real hot-spots. They are 'hot' precisely because they are politically complex. And on many of these complexities, the EU and the US simply do not see eye-to-eye – neither on the problem, nor on the solution.

As to the 'global war on terrorism', the European response consisted, at one level, in denying the very concept of 'war'. For Sir Michael Howard, the formulation of that concept was a 'terrible and irrevocable error [. . .]

to declare that one is at war is immediately to create a war psychosis that may be totally counter-productive for the objective being sought'.[10] This is not to say that the two sides of the Atlantic are not cooperating to fight the terrorists. At a meeting in Dromoland Castle, Ireland, on 26 June 2004, EU and US leaders issued the EU-US Declaration on Combating Terrorism. There is no doubt that EU–US cooperation on counter-terrorism has been substantial and growing ever since 9/11. Intelligence sharing has been more intense and more substantial than ever before – and this has included crucial Franco-US exchanges. Cooperation between law-enforcement agencies and prosecutors has been massively stepped up. On 25 June 2003, the two sides concluded an Extradition and Mutual Legal Assistance Agreement facilitating extradition for many more offences than previously. Despite serious European misgivings, agreement was reached in May 2004 on communication of Passenger Name Records (PNRs) in connection with international travel. This agreement, however, was annulled by the European Court of Justice in May 2006. In September 2004, wide-ranging agreements were reached on the safety of container transport (the Container Security Initiative), including extensive customs cooperation and the facility for US officials to check container cargoes in European ports. Joint US–EU investigative teams were being implemented. A wide-reaching Policy Dialogue on Border and Transport Security was attempting to narrow the gap on issues such as sky-marshals and biometric data. Substantial legal and banking cooperation was agreed on countering terrorist financing. This amounted to a substantial package of agreements, many of which would have been virtually unthinkable five years previously. However, agreement was largely on the administrative, legal and technical aspects of counter-terrorism. The part to be played by an organisation such as NATO remains very unclear.

The Alliance's 2004 *Military Concept for Defence against Terrorism* states the Alliance's readiness

> to act against terrorist attacks, or the threat of such attacks, directed from abroad against our populations, territory, infrastructure and forces; to provide assistance to national authorities in dealing with the consequences of terrorist attacks; to support operations by the European Union or other international organisations or coalitions involving Allies; and to deploy forces as and where required to carry out such missions.

It is not hard to see how a major military alliance *could* be involved in such activities: neutralising terrorist training camps, keeping open shipping lanes, deploying forces to maintain order after a terrorist attack. But in what way are these activities specific to NATO? In short, if an Alliance is defined by the threat against which it is created, it remains very unclear how or indeed whether 'terrorism' can replace 'communism' as the glue holding the structure together.

The NATO Response Force (NRF) of 17,000 troops reached initial operational capability in October 2004 and was theoretically ready to take on missions anywhere in the world. However, to date, it has not proved possible to find it a specific mission. One of the objectives of the NRF was to offer troops from European member states the prospect of training and deployment alongside US forces in order to enhance capabilities and improve interoperability. But American inputs were minimal and NRF rotations planned through to June 2007 remain overwhelmingly European in composition. These same finite forces may also be required for duty in the growing European military structures. The EU is planning, by 2007, to field up to fifteen 'battle-groups' for peace-keeping operations up to 6,000 kilometres from Europe. As the EU mission Artemis in the Democratic Republic of Congo demonstrated in June 2003, it is much easier for the Europeans to reach political consensus on a mission than it currently is for NATO. Moreover, there is no clarity about the division of labour between these European forces and the NRF. In short, the politics of NATO's operationality has become hostage to the overall state of the transatlantic relationship.

Matters improved somewhat under the second Bush administration. In her speech to Sciences-Po in Paris on 8 February 2005, Secretary of State Condoleezza Rice struck a conciliatory note, insisting that partnership with Europe was essential to the new US focus on the triumph of 'freedom', stating unambiguously that 'America has everything to gain from having a stronger Europe as a partner', and quoting from President Bush's inaugural address: 'All that we seek to achieve in the world requires that America and Europe remain close partners.' These were words all Europeans had hoped to hear. There was, however, scarcely a word about how NATO fitted into the overall picture. There is no question that the old Atlantic Alliance of the Cold War has run its course. Few in either Europe or the US would question that the two sides require a new, effective and clear-sighted strategic partnership in a turbulent world. In terms of trade and investment, the partners are more interdependent than any two entities have been in the history of the world. But to build a true partnership requires dialogue, listening and genuine mutual respect. It requires a willingness to work hard at identifying common interests and agreeing on how to defend them in common. The future of NATO as an institution hinges crucially on whether the coming years can usher in such a climate.

The forward march of ESDP

Meanwhile, the European Union's ESDP has taken large strides on its own. At the European Council meeting on 17 June 2004, Headline Goal 2010 (HG 2010) was adopted, committing the Union 'to be able by 2010 to respond to a crisis with rapid and decisive action applying a fully coherent approach to the whole spectrum of crisis management operations covered by the Treaty on the European Union'. Interoperability, deployability and sustainability are at the heart of the project and the member states have identified an indicative list of specific milestones within the 2010 horizon, including the establishment of the European Defence Agency (EDA) by the end of 2004; the implementation of an EU strategic lift joint coordination by 2005; the ability by 2007 to deploy force packages at high readiness broadly based on the EU 'battle-groups' concept; the availability of an EU aircraft carrier by 2008; and 'appropriate compatibility and network linkage of all communications equipment and assets' by 2010. HG 2010, by focusing on small, rapidly deployable units capable of high-intensity warfare, shifted the objective from quantity to quality. It also resolved (at least partially) the contradiction between a Kosovo-style capability and the requirements of the 'war on terrorism'. The newly created battle-groups, of which up to fifteen are projected for 2007, can be used for both types of operation. By 2005, the EU was beginning to look like an increasingly credible potential military actor. The battle-groups model was inspired by the first ever experiences of the EU in armed combat, which began in 2003.

On 31 March 2003, the EU launched its first military operation – a peace-keeping mission in the Former Yugoslav Republic of Macedonia (FYROM), taking over from a NATO force. Operation Concordia used NATO planning under the 'Berlin Plus' procedures. It deployed 357 troops (from all EU states except Ireland and Denmark, and from fourteen additional nations – an average of thirteen troops per participating member state) into a small mountainous country, successfully keeping the peace between bands of lightly armed irregulars and the Macedonian 'army' which boasts a defence budget less than half that of Luxembourg. This was an operation high in political symbolism and modest in terms of military footprint. By the end of the mission, it was clear that the biggest problem in Macedonia was no longer armed conflict but criminality – hence Concordia was succeeded on 15 December 2003 by an EU police operation: Proxima. Concordia's primary value was that it allowed the EU to test its recently agreed procedures covering every aspect of the mounting of a military operation – however modest – from command and control through use of force policy, to issues such as logistics, financing and legal arrangements and memoranda of understanding with host nations.

From June to September 2003, the EU launched its first ever autonomous operation outside of the NATO framework. Operation Artemis, in the Democratic Republic of Congo, offers even richer lessons about EU capabilities than Concordia. The initial assessment suggests that the mission, which involved rapid force projection to a distance of 6,500 kilometres into unknown and hostile terrain, was a success. France was the 'framework nation', supplying 1,785 of the 2,200 troops deployed. Sixteen other 'troop contributing nations' (TCNs) were involved, providing strategic air lift (Germany, Greece, United Kingdom, Brazil and Canada), engineers (UK), helicopters (South Africa) and special forces (Sweden). Operational planning was conducted from the French Centre de Planification et de Conduite des Opérations (CPCO) at Creil, to which were seconded officers from thirteen other countries, thus demonstrating the potential for multinationalisation of a national headquarters. The operation was exemplified by rapid deployment (seven days after UNSC Resolution 1484 on 30 May), a single command structure, appropriately trained forces, clear rules of engagement, good incorporation of multinational elements, excellent inter-service co-operation, and adequate communications. NATO procedures were used throughout. Artemis demonstrated conclusively that the EU can undertake a peace-keeping operation, and on a significant scale, even at some distance from Europe. This may not yet equate to Kagan's notion of 'power'. But it was moving the EU clearly out of the realm of 'weakness'.

The transfer (from NATO to the EU) of responsibility for the Stabilisation Force (SFOR) in Bosnia-Herzegovina (BiH), in December 2004 (Operation Althea), represented an even greater test of the EU's military muscle. The initial NATO force deployed in BiH (IFOR, December 1995) involved some 60,000 troops. This was scaled down constantly to a January 2003 total of 12,000. Projections for 2005 foresaw a further reduction to about 7,000 troops centred on ten battle-groups of around 750 soldiers each. Over 80 per cent of the troops in NATO's SFOR were already from EU member states. Operation Althea was the EU's most ambitious military mission to date whose long-term objective was 'a stable, viable, peaceful and multiethnic BiH, cooperating peacefully with its neighbours and irreversibly on track towards EU membership'.[11] It was preceded by a range of indispensable strategic and political developments.

The European Security Strategy (ESS), approved by the European Council on 12 December 2003,[12] aimed to harmonise the different views of the current and future member states without falling into lowest common denominator rhetoric. The paper inevitably constitutes something of a compromise between different cultures and approaches among the EU's member states. The first section deals with the global security environment and gives recognition to the mixed perceptions of globalisation that exist. It pays greater attention to the root causes of poverty and global suffering than its US equivalent.[13] It identifies five key threats: terrorism, weapons of mass destruction (WMD), failed states, organised crime, and regional conflicts. The document has been criticised in some quarters for its alleged alignment, via this focus on threats, with US security policy. The EU document is nevertheless more nuanced than its US equivalent, stressing the 'complex' causes behind contemporary international terrorism and recalling the destabilising effects of regional conflicts such as Kashmir, the Great Lakes and the Korean peninsula, all of which feed into the cycle of terrorism, WMD, state failure and even international criminality. Nonetheless, it is unequivocal in stating that the EU faces the same challenges as the United States.

The second section outlines the EU's 'strategic objectives'. Two features are stressed: that 'the first line of defence will often be abroad' – via conflict prevention; and that none of the new threats is 'purely military' or manageable through purely military means. The strategic objectives rest on two main pillars: building security in the European region, and creating a viable new international order. The former is absent from US policy, the latter only fleetingly entertained. The EU document is strong in its assertion of a commitment to upholding and developing international law and in recognising the UN as the main source of international legitimacy. However, the most innovative aspect of this section is the new emphasis on using the EU's powerful trade and development policies in a conditional and targeted way.

The final section addresses the policy implications for the EU. The EU needs to be 'more active, more capable and more coherent'. One of the boldest statements of the document (which guaranteed applause in the United States) is the need to develop a strategic culture that fosters 'early, rapid and, where necessary, robust intervention'. The Strategy, it is claimed, will contribute 'to an effective multilateral system leading to a fairer, safer and more united world'.

The Constitutional Treaty of August 2004 called for a range of new developments in the field of ESDP. The fact that the Treaty failed in the summer of 2005 did little to slow down these developments. The creation of a European Defence Agency subject to the authority of the Council was already an established fact. Armaments cooperation had hitherto taken place rigorously outside the EU framework. Two main reasons lay behind this belated decision to change tack. The first was the relative failure of previous attempts to coordinate procurement and armaments cooperation. The second was the accelerating reality of ESDP and the concurrent perceived need to link capabilities to armaments production. In early 2004, an Agency Establishment Team set about clarifying its objectives and role and narrowed down four basic purposes:

- to work for a more comprehensive and systematic approach to defining and meeting ESDP's *capability* needs;
- to promote *equipment collaboration*, both to contribute to defence capabilities and to foster further restructuring of European defence industries;
- to encourage the widening and deepening of regulatory approaches and the achievement of a European defence *equipment market*;
- to promote defence-relevant *research and technology* (R&T), 'pursuing collaborative use of national defence R&T funds' and 'leveraging other funding sources, including those for dual use or security-related research'.

The EDA is guided by a Steering Board meeting at the level of Defence Ministers, nominally headed by the HR-CFSP (later Union Minister for Foreign Affairs) and managed by a Chief Executive. It offers the first real opportunity for the EU to bring its defence planning, military capability objectives and armaments coordination in line with the urgent tasks it is facing on the ground. The EU governments are poised to take a major step forward towards more rational armaments and defence planning. The dynamics of ESDP suggest that they will progressively situate their national plans within a European framework.

Another breakthrough came in 2003 in the field of operational planning. This was a contentious issue which had, for several years, pitted the UK against France. Operational planning is the *sine qua non* of any military mission. Paris had always been keen to develop autonomous EU operational planning capabilities, but London had resisted, arguing that this was an expensive duplication of an already existing NATO capability. However, in summer 2003, as Tony Blair sought to mend fences with his European partners, a compromise was arrived at involving three distinct operational planning facilities. For EU operations under 'Berlin Plus', a dedicated EU unit has been attached to NATO at SHAPE. For most 'EU-only' operations, including most battle-group missions, an appropriate national headquarters will be multinationalised. But for certain EU-only operations, particularly those involving combined civil and military dimensions, a dedicated (and autonomous) EU planning cell is being developed at ESDP headquarters in the rue Cortenberg in Brussels – more evidence of an embryonic EU 'power'. Another potential breakthrough in 2004 came in the field of intelligence gathering.The agreement between the EU and the US for the coordination of the former's Galileo satellite navigation system with the US's GPS system potentially constitutes a major step forward on the road to EU autonomy in intelligence gathering.

The Constitutional Treaty also proposed a new Union Minister for Foreign Affairs (UMFA) combining the current responsibilities of both the HR-CFSP and the Commissioner for External Relations, thus having one foot in the Council and one (as Vice-President) in the Commission. This would allow the Minister to coordinate the two main thrusts of the EU's external policy: security and overseas aid. The post-holder, elected for a five-year term, was intended to replace the previous semestrial rotating presidency and to preside over a 'European External Action Service', which was intended to be introduced within one year after entry into force of the Treaty.[14] Although this development was temporarily halted by the constitutional train-wreck of summer 2005, most commentators on that crisis identified the new ministerial post as the most important item to be salvaged from the wreckage itself. The focus over the next few years will be on finding a politically acceptable way of achieving that aim.

Conclusion

Two powerful forces have projected the EU towards some form of ESDP. The first was the result of exogenous pressures from the US and other parts of the world. Although US pressure was a major midwife of ESDP, US reactions to the new infant have been varied and contradictory. The initial (semi-official) reaction was of the 'Yes, but . . .' variety. ESDP would be acceptable to Washington if it proved to be what the St Malo Declaration said it intended to be: a contribution to a revitalised transatlantic alliance. For this to happen, certain conditions must be met, including Secretary of State Madeleine Albright's instant strictures against NATO 'decoupling, duplication and discrimination'.[15] This position has ebbed and flowed in the reactions of administration spokespersons with the regularity of the tides. It has alternated with the expression of serious opposition to the entire ESDP project. This has assumed two contradictory forms. The opposition is motivated by a belief either that ESDP will in fact emerge as a rival to NATO, or that it will prove to be such an incompetent force that it will merely aggravate Alliance tensions. As ESDP began to emerge as a viable military actor in 2003–4, this latter fear was silenced and the 'debate' on ESDP featured the two former attitudes. Official Pentagon and State Department spokespersons, faced with a US military which was vastly overstretched in late 2004, began reiterating the initial ESDP mantras: European military capacity was to be welcomed because, if EU–US political tensions could be resolved, the European force could relieve the US armed forces. However, other conservative commentators, scrutinising some of the debates within the European Convention of 2002–3, were forced to the conclusion that the 'political integration of the EU presents the greatest challenge to continuing US influence in Europe since World War II, and US policy must begin to adapt accordingly'.[16] The US administration is faced with an 'agonising re-appraisal' of its entire attitude towards European unity. Meanwhile ESDP continues to thrive.

Endogenously, the internal developments of the European integration process have had a marked effect on this new policy area. Much has been accomplished very rapidly in a period of five short years. Yet any overall evaluation of European military capacity has to be set within a very clear timeframe. Over the next five years (the short term), the EU will need to absorb the lessons of Artemis, Althea and other such operations, concentrate on plugging the gaps in strategic assets, develop genuinely integrated operational capacity and perfect command, logistics and communications procedures. This will limit actual operations to the type and style of those we have already witnessed and are about to witness. During this period also, however, the EU will need to plan procurement projects for the medium term (the following fifteen years). This will necessitate tough political decisions about the ultimate size, scale and style of EU military ambitions. How far down the road towards US-style network-centric warfare will the Union wish to go? How many new-generation platforms and other strategic systems will it require – for what purposes? How far afield does it anticipate intervening? In parallel, the Union will have to develop a holistic Strategic Concept along the lines of the proposals discussed above. Without clear guidelines as to its ultimate objectives and purpose, progress will be stalled. In the longer term, only two factors prevent the EU from developing genuinely autonomous and seriously credible military muscle: its ability to cooperate and to integrate, and the political will to implement its decisions and to act robustly in support of the values and interests outlined in its evolving security strategy. That is the internal challenge facing the next generation of Europeans.

Notes

1 Richard, E. Rupp, *NATO After 9/11: An Alliance in Decline* (New York: Palgrave, 2005).
2 Sten Rynning, *Nato Renewed: The Power and Purpose of Transatlantic Cooperation* (London: Palgrave, 2005).
3 IISS (International Institute for Strategic Studies), *The US Global Posture Review, Strategic Comments* (2004).
4 Jean-Yves Haine, *From Laeken to Copenhagen: European Defence. Core Documents Volume III* (Paris: EU-ISS – Chaillot Paper 57 2003), pp. 178–80.
5 'The Parties agree that an armed attack against one or more of them [. . .] shall be considered an attack against them all.'
6 Jolyon Howorth and John Keeler (eds) *Defending Europe: The EU, NATO and the Quest for European Autonomy* (New York: Palgrave, 2003).
7 Turkey, Norway and Iceland – joined in April 1999 by the Czech Republic, Hungary and Poland.
8 Haine, *From Laeken to Copenhagen*, pp. 178–9.
9 Senator Lugar had declared, in 1993, that NATO must 'go out of area or out of business'. Article 6 of the North Atlantic Treaty had limited NATO's area of activities to the territories of its member states.
10 Sir Michael Howard, 'What's in a Name? How to Fight Terrorism', *Foreign Affairs*, 81/1, January/February, 2002.
11 http://ue.eu.int/cms3_fo/showPage.asp?id = 745&lang = en&mode = g
12 [Solana Strategy Paper, 2003] European Council, *A Secure Europe in a Better World*, Brussels, 12 December 2003, accessed May 2004 at: http://ue.eu.int/uedocs/cmsUpload/78367.pdf
13 [US National Security Strategy, 2002], *The National Security Strategy of the United States*: http://www.whitehouse.gov/nsc/nss.html
14 S. Duke, 'Preparing for European Diplomacy?' *Journal of Common Market Studies*, 40/5, 2002.
15 M. Albright, 'The Right Balance Will Secure NATO's Future', *Financial Times*, 7 December 1998.
16 J. Cimbalo, 'Saving NATO from Europe', *Foreign Affairs*, 83/6, 2004; and J. Hulsman, 'A Conservative Vision for US Policy Towards Europe', *Heritage Foundation Backgrounder* 1803 (2004).

9

THE TRANSATLANTIC ALLIANCE IN THE IRAQ CRISIS

Saki Ruth Dockrill[1]

In an earlier study on the transatlantic alliance in the aftermath of the terrorist attacks on September 11, 2001 (9/11), I argued that 9/11 was a wake-up call for the globalisation of security beyond the transatlantic region. It characterised the growing differences within the alliance in its approach to global terrorism and America's dismissive attitudes towards NATO. That 9/11 threatened the credibility of the transatlantic alliance was the conclusion of that study.[2] In this chapter, the focus will be on the transatlantic crisis over the war against Iraq, and examining the scope and nature of the rift between Europe and the US. At one level, the argument was concerned with the wisdom of going to war in the spring of 2003, although all the major capitals in Europe agreed that Saddam Hussein's Iraq was a serious concern to the security of the Middle East. At a wider level, the debate on both sides of the Atlantic was centred on the question of trust and confidence between the Europeans and the Americans. Many critics asked whether the emerging differences of values and beliefs in both societies might have been at the root of the transatlantic rift. Between the autumn of 2002 and the spring of 2003, the gulf in the transatlantic alliance appeared at times so fundamental and deep that many scholars and commentators believed that the alliance was in terminal decline, or even already dead. Alternatively it was heading for an 'amicable divorce'. Since then much ink has been spilt over the subject and the current prognosis of the alliance still remains negative. The transatlantic alliance, Charles Kupchan wrote, 'is gone for good'. Europeans echo this sentiment by describing Europe as 'Post-Atlantic societies'.[3]

This chapter examines whether a major paradigm shift has occurred in America's relations with Europe, a concept first coined by a scientist, Thomas Kuhn, in his book, *The Structure of Scientific Revolutions*.[4] It will examine whether America's perception of the transatlantic alliance has finally broken with the tradition which had been established during the Cold War years, and whether it no longer constitutes a major factor in US foreign policy. If this is the case, what was responsible for this change? In answer to this, the chapter examines the ways in which the transatlantic alliance became divided over the Iraq crisis, first, between the political elites, and second within both societies in the form of anti-Europeanism and anti-Americanism. In the latter part of the chapter, the rift will be analysed at three levels. From the historical and cultural perspectives, it will examine how the United States came to acquire the status of the world's most hegemonic power, and what ideas and beliefs have helped to shape the United States into what it has now become. The second level of analysis is concerned with structural changes. The transatlantic alliance has survived the end of the Cold War, but for the US the Cold War proved to be a long period of constraint. During that war, the US had to maintain America's alliance with Europe as essential to winning the Cold War. For the Europeans, the Cold War meant that their security was dependent upon the nuclear deterrence extended from the US, and its end reduced Europe's security dependency on the US. Robert Cooper, a senior British diplomat, argues that the European states have now entered into the post-modern international system. How far have the structural changes since the end of the Cold War contributed to the current transatlantic rift? Finally, while human agency is constrained by structural factors (such as institutional and national interests and their limits), it is also true that decision makers are often faced with a range of choices.[5] In this context, a majority of Europeans tend to see the Bush administration as representing American neo-conservatism, and hence at the core of their current problems with the United States. The chapter will investigate the factor of human agency to see whether the personalities and ideas unique to the Bush administration have caused the transatlantic crisis over Iraq.

The road to the crisis

Prior to the 9/11 terrorist attacks, the growing military gap between Europe and the United States had already become a major issue. Washington initially regarded post-Cold War conflicts in Bosnia and Kosovo as being within Europe's sphere of responsibility, but as it turned out, US military power and leadership became essential to the defusing of these crises. For some time, the Pentagon dismissed NATO as militarily irrelevant and the experience of Kosovo only reinforced the view in Washington that co-operation with Europeans in military campaigns was becoming less and less attractive. Writing in 2000, Condoleezza Rice (who was soon to take up the post of Bush's National Security Adviser) stated that NATO 'will mean nothing to anyone if the organisation is no longer militarily capable and it is unclear about its mission'.[6] During the post-Cold War years, the US had encouraged NATO to take on more security roles beyond Europe, without much success, but with the shock of 9/11, the Europeans were galvanised into doing just that. They offered any assistance which might be required by their long-term ally, the United States, at its most vulnerable, but NATO's invocation of the Article 5 security guarantee was politely declined by Washington. Instead, the new Republican administration under George W. Bush preferred a 'shifting alliance' which would be 'opportunistic, temporary, constantly shifting'. Defence Secretary Donald Rumsfeld maintained that the mission should determine the coalition, and not the other way round. Undeterred, however, many European states offered arms and manpower to the American-led coalition of the 'willing' in the anti-al Qaeda campaign in Afghanistan. This was, too, regarded by the Pentagon as more of a nuisance than an asset to their military campaign. In the meantime, the US mounted its successful military operation in Afghanistan, which boosted further the Pentagon's 'can-do spirit'.

No sooner was the Afghan campaign over than the cracks became more apparent in the unity of the alliance: the Bush administration became determined to take on 'the war on terror' beyond Afghanistan, against the 'Axis of Evil' (Iraq, Iran and North Korea), and especially against Saddam Hussein's Iraq. In the summer of 2002 Bush also launched the strategy of pre-emption ostensibly against Iraq in order to deal with American vulnerability exposed by 9/11. The salience of the 'Bush Doctrine' is to emphasise assertive US leadership, combined with its willingness to resort to the use of America's military power.[7] This assertive or militant American stance reinforced Europe's scepticism about Washington's intentions. Added to this was a series of unilateralist approaches to the international community, which included withdrawal from the Anti-Ballistic Missile Treaty, the imposition of high protective tariffs on European, Russian and Asian steel, and America's opposition to the creation of the International Criminal Court (ICC). In the field of global terrorism, it has become clear that Europeans are more inclined to choose a softer long-term policy to win the hearts and minds campaigns against terrorists rather than resorting to military force, but it is the issue of Iraq that has divided the transatlantic alliance most.

Iraq had been an outstanding issue in Washington and to a lesser extent in London ever since the first Gulf War of 1990–1. The experience of this war led the Clinton administration to adopt the Theatre Missile Defence system by abandoning the idea of a multilateral approach to the issue of the proliferation of weapons of mass destruction (WMD). Saddam continued to behave in a threatening manner to Iraq's neighbouring countries, supported a plot in 1993 to kill George H.W. Bush, and deluded and frustrated UN inspectors who were searching for evidence of Saddam's exploitation of unconventional weapons. During the post-Gulf War years, the US continued almost daily air strikes in the so-called slow-motion war in Iraq, and the press eventually lost interest in it. In 1998, the US and British in cooperation mounted Operation Desert Fox in order to destroy Baghdad's chemical and biological stocks. The United States hoped that anti-Saddam resistance forces would topple Saddam from power, but neither the strategy of 'roll back to change the Baghdad regime' nor that 'to keep Saddam Hussein in his box' worked, and Saddam remained in power.[8]

After 9/11, the Bush government skilfully put the ongoing security concerns into the one single box of national security containing the 'trinity of evil'[9] – terrorism, the WMD threat, and the rogue states (the last of which came to be called after January 2002 'the Axis of Evil'). The 9/11 tragedy served to change the balance of power between the President, Congress (which was in any case controlled by the Republicans) and other opinion-forming and interest groups, since the 'politics of US national security is always a reflection of the sense of national danger'.[10] The Cold War was sustainable because of the sense of national danger from Soviet or other Communist nuclear attack. After 9/11, national security is now back high on the global agenda of the US under the firm control of the White House. Saddam's potential as the supplier of WMD to terrorist groups like al Qaeda strengthened Washington's anxiety to change his regime and remove WMD from Iraq. In Britain, which also put Iraq high on its security agenda, Tony Blair and his close associates had held intensive discussions with the Clinton and Bush administrations over Iraq for some time. Britain however attached greater importance than the US to the securing of UN sanctions for the use of force, and it was only after a further, and ultimately unsuccessful, Anglo-US pursuit of a second resolution authorising military action against Iraq in March that the two powers decided to resort to war regardless of the UN.

The Anglo-US position on Iraq was viewed sympathetically by the pro-Atlanticist governments of Spain, Italy, the Netherlands, Denmark, Portugal and Ireland, as well as by most of the Eastern European countries. On the other hand, the governments of France, Germany and Belgium led the anti-war Euro-opposition group. France, in the 1970s, had contributed, together with the Soviet Union, to the strengthening of Saddam's Iraq. Jacques Chirac, then Prime Minister, had overseen an agreement with Saddam, which led to France's sale of enriched nuclear material and reactors, as well as the transfer of nuclear technology by training Iraqi scientists and technicians. After the Gulf War, France had voiced its concern about the limited utility of the sanctions imposed by the allies, and since the mid-1990s, France had chosen, instead, to engage with Iraq economically, in the form of the 'oil-for-food programme' worth $1 billion in 2002.

Washington was particularly irritated by France's last-minute effort to squash the planned Anglo-American second UN resolution. On 10 March 2003, Chirac (now the French President) described France's position in no uncertain terms, insisting that it would 'veto any new ultimatum to Iraq whatever the circumstances'. The French position was supported in the UN by Russia and China (two permanent Security Council members) and also by Germany and Syria. In the summer and autumn of 2002, Washington was led to believe that France's position was coming closer to that of the US: France had a discussion with Washington about its potential military contribution, including 15,000 troops and 100 airplanes, in the event of war, while the French Foreign Minister, Dominique de Villepin, dismissed Saddam's regime as one that 'for years had defied the international rules defined by the UN'. While at every point France expressed its opposition to 'unilateral action',[11] the US mistakenly upheld its conventional perception of France as 'a troublesome but ultimately reliable ally that could be counted upon in times of crisis'.[12]

The US also miscalculated the intentions of Germany, another heavyweight in Europe. The German government's well-known pro-Atlanticism normally enjoyed strong support from the German population. The country had been unified with the blessing of George Bush's father in 1990. Berlin, however, like Paris, demonstrated to Washington that it was prepared to sacrifice its traditional Atlanticism when the circumstances warranted it. The German Chancellor, Gerhard Schröder, sided with the majority of the German people, and 'chose to campaign shamelessly and relentlessly against the US and a possible war in Iraq' in the 2002 elections. This angered the Bush administration. Only a few months earlier, the Chancellor had expressed his utmost understanding of the US's position on Iraq, and promised Bush that he would not use Iraq during his election campaign. Not only did Schröder decide to win popular votes by expressing his opposition to the use of force against Iraq, but he also cooperated with Chirac in January 2003 to consolidate the Franco-German opposition to a possible war against Iraq. During the celebration of the fortieth anniversary of the Elysée Treaty, the two leaders announced their agreement that only the UN Security Council could authorise the use of force against Iraq, but that neither Paris nor Berlin believed that the West had exhausted the non-military means of disarming Saddam.[13]

The months leading up to the war in 2003 had been extremely difficult for all concerned: Iraq proved to be more cooperative over arms inspections than anticipated, thus strengthening the hands of those major Western European powers who opposed Bush's militant policy towards it. There was also a dispute over procedure. President Bush, the American public, and Congress were all satisfied with UN resolution 1441, which had been adopted unanimously by the Security Council in the autumn of 2002. The resolution warned Iraq that it would face 'serious consequences' in the event of its non-compliance with the UN inspectors. Saddam's 12,000-page report which arrived at the UN in December 2002 fell short of such full cooperation, thereby in American eyes justifying military action. However resolution 1441 failed to establish a consensus needed for such action in the UN, and after January, the controversy surrounding 1441 widened the gap between the pro-war and the anti-war factions within the transatlantic alliance. Despite Bush's assertive unilateralism, he counted on Britain's military assistance, but Tony Blair, faced with a hostile media, a sceptical public, and anti-war Labour backbenchers, needed a second resolution in order to justify war. Bush realised that without London, 'he could be close to going alone', which would make America's action seem even more imperial. At the end of January, Bush, as requested by Blair, agreed to back a second resolution. At that point, some observers wondered whether war was less likely now that the US was apparently pursuing a 'multilateral path' in the UN.[14]

The US adopted the position of negotiating with the UN, while preparing for war against Iraq. The deployment of American forces in the Gulf area was continuing at full speed despite the stagnated debate in the UN over Iraq. On 3 January 2003, Bush spoke publicly to a US army audience at Fort Hood in Texas stating that 'a war in Iraq' would be 'not to conquer but to liberate'.[15] Meanwhile UN inspectors were still trying to discover WMD in Iraq, and on 7 March 2003, the UN team asked for more time (probably several months) in which to complete their task. Throughout this period, however, the US underestimated the extent of Western Europe's opposition to the war. While Washington had declined NATO's cooperation in the Afghan campaign in 2002, it assumed that in the case of the Iraq operation, the US would need and would secure NATO's military and technical assistance. Washington particularly wanted Turkey's cooperation, which would allow US troops to move into northern Iraq. However, Germany, France, Belgium and Luxembourg blocked the American proposal calling for NATO's defence of Turkey in the event of a war against Iraq, as the leaders of these countries felt that 'it was premature and overly broad and would appear to commit NATO to supporting a war the Security Council had not yet approved'.[16]

The angry debates in Brussels over Iraq in February 2003 prompted a former American ambassador to NATO to remark that it was 'the worst split I've ever seen in NATO'.[17] Eventually there was a compromise: the US scaled down its request, abandoning the use of European forces to be replaced by allied forces. Belgium and Germany, for their part, agreed to prepare NATO for the defence of Ankara. Another twist came on 1 March 2003 when the Turkish parliament rejected a bill authorising America's access to bases in Turkey in the event of war. This was despite the fact that Bush had tried to lure Turkey with an offer of a $26 billion aid package. Both the Turkish military establishment and the general public had, however, been unenthusiastic about the prospect of a war in Iraq for some time, fearing that it might promote the idea of Kurdish autonomy in northern Iraq, which would in turn incite Kurdish separatism in Turkey.

This was not the end of setbacks for the allies prior to the war. Britain's special intelligence dossier on Iraq was discredited by a revelation in February 2003 that eleven out of nineteen pages in the document had been directly copied from an American graduate student's essay, which had been written nearly twelve years before. Moreover, before this revelation, the US Secretary of State, Colin Powell, had publicly expressed his admiration for the British dossier at the UN Security Council. In his State of the Union speech in January 2003, President Bush also mentioned that the British government had 'learned that Saddam Hussein recently sought significant quantities of uranium from Africa', but in March, the International Atomic Energy Authority told the UN that Niger, a former French colony in Africa, had not provided Saddam with uranium, contrary to the evidence claimed by the British intelligence community. Apparently the document on which this claim was based had been forged.[18] These blunders made the Anglo-US alliance look like a group of boy scouts. Chirac's 10 March statement on the use of France's veto was the final straw for the Anglo-American leaders, as there would be no point in presenting a second resolution only to have it vetoed by France. On 17 March, when the proposed second resolution was withdrawn from the UN Security Council, Sir Jeremy Greenstock, the British Ambassador to the UN, complained that 'One country in particular has underlined its intention to veto any ultimatum.'[19] The war was launched on 20 March and the transatlantic alliance was now openly divided. The legitimacy of the war was not fully established in the UN nor in the international community.

Anti-Europeanism and anti-Americanism

Nearly 70 per cent of Americans supported the Iraq war while more than 80 per cent of Europeans opposed it.[20] Bush's familiar phrase, 'If you are not with us you are against us', echoed American sentiments at large: between February 2002 and March 2003, Americans who had previously viewed the French favourably declined from 79 to 29 per cent, and the Germans from 83 to 44 per cent.[21] There followed in the US a huge anti-French and anti-German backlash, calling for boycotts of French and German goods. France and Germany were branded as the 'Axis of Weasel'. In the cafeteria at the US House of Representatives, 'French fries' were renamed 'Freedom fries'.[22] Donald Rumsfeld dismissively called France and Germany 'Old Europe', living back in the twentieth century, and ungrateful and cowardly allies. American criticism also extended to Belgium, which had tried to block the US proposal on the defence of Turkey in NATO. The *Wall Street Journal* published a scathing comment on the Belgian armed forces, which could not fight properly because they had 'hundreds of hairdressers, musicians and other non-combatant personnel at their disposal'.[23] In this acrimonious debate against 'Old Europe' as opposed to 'New Europe', the US Senate put forward a resolution commending those pro-American European allies for standing alongside the US on Iraq, an implied criticism of France and Germany and other countries (Belgium, Greece, Austria, Sweden and Finland), which did not. Given their turbulent histories, the former socialist Eastern European countries as well as the Baltic states regarded the United States (and not Germany or France) as their ultimate guarantor, and were decidedly Atlanticist in their orientation. The Republican Senator from Arizona, John McCain, once regarded as the leader of the neo-conservatives, was pleased to state that 'the majority of Europe's democracies have spoken, and their message couldn't be clearer: France and Germany do not speak for Europe'.[24]

The difficulty with America's New and Old European tactics was the fact that a large majority of Western Europeans were supporting the stance taken by the French and German governments. Many streets in Western European capitals saw anti-American fervour running high. Interestingly, anti-war demonstrations were more vocal in the countries whose governments supported the US's position on Iraq, such as Britain, Italy and Spain. On 15 February, London saw the largest public protest thus far, consisting of nearly one million people. Similar numbers of anti-war demonstrators were recorded in Madrid, while nearly two million people took to the streets in Rome. However, a poll taken in mid-March indicated that the bulk of the British people would support war 'if it had UN backing', but only 15 per cent answered positively in support of war regardless. These anti-war demonstrations reflected a growing anti-Americanism which became 'a regular feature of the political landscape' in Europe. Americans were viewed as militant, arrogant, unilateralist and vulgar. The strong and simple language emanating from the Bush administration merely boosted anti-American sentiments.

The evolution of American ideas and the world

The question remains whether the crisis over the Iraq war was merely a symptom and not a cause of the transatlantic rift. One striking example of the former came from Robert Kagan, who insisted that 'Americans are from Mars and Europeans are from Venus: they agree on little and understand one another less and less.'[25] The trouble with the Europeans, according to Kagan, is their inability to develop their military power sufficiently to play the global role to which they aspire. However, 9/11 demonstrates that the world remains a dangerous place, and it is the US, by virtue of its superior military might, that has the power to deal with, and fix, problems. Kagan asserts that Europeans have now come out of 'the Hobbesian world of anarchy into the Kantian world of perpetual peace', or into 'paradise'.[26] Other Americans, including John Lewis Gaddis, Melvyn P. Leffler, Eliot Cohen and Jonathan Monten, all argue that the Bush administration's policy does not represent anything particularly revolutionary, and, if anything, Americans could now behave more like the true Americans of old.[27] This is because their sense of insecurity has increased after 9/11, because of fewer restraints upon the conduct of American foreign policy after the end of the Cold War, and because the US has now achieved the status of 'hyper puissance'.

The United States has always occupied a unique position in the history of great powers because of its historical origins, geopolitics, capabilities and political beliefs. Born out of revolution and war, by sheer determination and resolve, the United States obtained its independence from Britain. Hence, the United States regarded itself as an anti-colonial and an anti-imperialist power, conscious of the limits of its power, and with a deep sense of 'vulnerability to external pressures in a world of scheming nations'.[28] These factors, combined with the United States' earlier experiences with the European imperial powers, codified its mental map in the form of isolationism, which was expressed in the early part of the twentieth century by neutrality and the avoidance of close entanglement with the European powers. The other side of isolationism was the rejection of foreign intervention in American affairs, and this mindset was epitomised in the Monroe Doctrine of 1823, which amounted to a declaration of the US's legitimate right to determine its own destiny on the American continent, as demonstrated by its frequent military interventions in Latin America in the twentieth century. From this, one school of thought suggests that American isolationism could in fact be more suitably termed as 'unilateralist'.

The sense of vulnerability to external threats and the rejection of European interventionism further encouraged America's incentive to pursue *bona fide* independence, and to achieve 'national greatness'. By the beginning of the twentieth century, the US had emerged as an expansive, dynamic and industrial power to be reckoned with in the world community. As the country became stronger and more united, it increased its impulse to disseminate its values to the outside world and to 'intervene in order to guide revolutions'.[29] Captain Alfred Thayer Mahan and Theodore Roosevelt were both leading enthusiasts for the Americanisation of the world. The Spanish–American war of 1898 was the catalyst for Mahan's imperial view of naval strategy, and resulted in the annexation of the Hawaiian islands and part of Samoa, and the seizure of the Philippine islands and Guam from Spain. The war led to the strengthening of American power in the Asia-Pacific region, and also over the Latin American states. The completion of the Panama Canal in 1914 enabled the US to exert a much wider control of sea communications from the Pacific through to the Caribbean.

The imperial thrust was replaced in the 1920s by an isolationist upsurge, although this did not constrain the expansion of the US economy worldwide, and by the early twentieth century, the US had become an economic superpower with the size of its industrial potential almost equivalent to the combined economic and industrial strengths of Britain, Germany, France, Russia and Japan. President Woodrow Wilson developed his idealistic views about the world order based on Jeffersonian values. Wilson's 'deep faith in the ultimate triumph of liberty' meant that he could be tolerant about foreign revolutions in the hope that the revolutionaries would eventually learn 'to preserve their self-control and the orderly processes of their governments'. On the other hand the same enthusiasm led the President to think ambitiously about remaking the world in America's image based on 'ordered liberty, free-trade, and international progress and stability'.[30]

While Wilson's cherished international organisation, the League of Nations, had never become sufficiently strong (mainly due to the absence of the US) or united to deter the Axis powers (Germany, Japan and Italy) from launching the Second World War, Franklin Roosevelt modified Wilson's unsuccessful internationalism into more realistic aspirations which could be acceptable to Congress as well as to the other great powers – Britain and the Soviet Union. The outcome was the creation of the United Nations. The new international organisation was built around America's ideas of national self-determination based on sovereign equality, but John Lewis Gaddis maintains that Roosevelt inserted into the Wilsonian project a 'cold-blooded, at times even brutal, calculation of who had power and how they might use it'.[31] Disillusioned by the weakness of continental Europe, the UN was an American attempt to replace the European concepts of spheres of influence and of a balance of power by the notion of collective security, but it also invested considerable power in the Security Council on which the US holds a permanent seat. The subsequent outbreak of the Cold War, and the domination of the superpowers in the international system meant that there was no 'sovereign equality' in the UN, and it became yet another sphere of Soviet–US confrontation. The US fought the Cold War with strategies of containment and deterrence, and it won that war. This has convinced many conservatives and liberals that the US's values and beliefs have successfully prevailed in the world.

Thus, American critics claim that the US had always been unilateralist and hegemonic and has always embraced a strategy of pre-emption to safeguard its own national interests. They point out that it is the Europeans who have changed over time: the Europeans were more willing in the past to resort to the use of force to resolve their rivalries, to exert hegemonic power through imperialism, and were certainly not always law abiding. In the case of the Suez crisis in 1956, argued Philip H. Gordon, the United States reacted in the same way as 'old Europe' has today. The Eisenhower administration counselled London and Paris against the use of force against Nasser's Egypt, and took the matter to the UN. The US's position was supported by the then Secretary-General, Dag Hammerskjold. The latter persuaded Britain and France to accept a ceasefire and sent a UN Emergency Force (the prototype of the current UN peace-keeping force) to replace the Anglo-French troops when the latter left Suez. Americans regard Europeans as hypocritical, as the latter now wish to be seen to be ardent supporters of the UN, multilateralist, and even pacifist, whereas they, too, at one time, put 'their faith in unilateralism and military force' to deal with intractable problems in the world.[32] Thus, the differences within the transatlantic alliance lie not so much in the cultural or historical origins of the US, but in the evolutionary changes which have taken place in the US and Europe.

Structural changes

The structural problems on both sides of the Atlantic are now familiar: the US has become the world's sole hegemonic power, while Europeans have become inward-looking, less militant, and less decisive. The US believes that Europe's clamour for multilateralism is a recent phenomenon to cover up their military weakness, as it 'lacks the capacity to undertake unilateral military actions, either individually or collectively as "Europe"'.[33] The assumption behind this argument is that if Europe had become as strong as the US militarily and economically, the current transatlantic rift would have been minimised. It might follow from this that Europe would therefore have agreed to the US's military solution in Iraq. Clearly this is a misleading conclusion. First, Europe was not united in response to the Iraq crisis. The majority of Europe's populations were against the war, and those governments which went to war against the strong tide of anti-Iraq war public opinion were punished, to varying degrees. For example, in the aftermath of the March bombings in Madrid in 2004, José Maria Aznar's government was voted out of power in favour of the Socialist opposition party. Although the main reason for his downfall was that the public felt cheated by Aznar's ready reference to the responsibility of the Basque armed group for the outrage, this was a country where 90 per cent of the Spanish population vehemently opposed war with Iraq. The incoming government was warmly supported in its decision to withdraw Spanish troops from Iraq. Similarly, Tony Blair's third election in May 2005 resulted in a substantial reduction of Labour's majority from 167 to 66. Blair's special relationship with Bush's war with Iraq was clearly a factor in explaining the dwindling of his popularity and that of his party. On the other hand, the French and German leaders, who led the strong anti-war sentiments in Europe, did not necessarily enhance their credibility as the leaders of Europe. Second, it was true that the end of the Cold War removed a 'coherent framework' from the Atlantic alliance – it had lost its common enemy. However, if the Cold War was understood to be a conflict of ideas, that is, Communism versus Capitalism, then in the age of global terrorism, both Europe and the US continue to face the same challenge from extremists, who are bitterly opposed to Western values and beliefs. Finally, the argument that the Europeans have lost their military ability to deal with international conflicts does not hold true either: the Europeans have not abandoned the use of force, as indicated recently in the military conflicts in Bosnia and Kosovo. Many European troops are still engaged in Afghanistan to restore order there. Nor was Europe's use of force strictly limited to humanitarian considerations, as France's (political) intervention in Rwanda bears witness. Jacques Chirac's France gave Washington the impression that it was willing to go along with the US during the Iraq debate in the UN and this also demonstrated that the Americans had not necessarily seen Europeans as having lost the will to fight in the event of a crisis.

There have certainly been structural changes in European politics. After having experienced two world wars in the first half of the twentieth century, Western Europeans have learned to resolve national rivalries through integration, rather than military conflict. In the age of growing pressure for decolonisation (a pressure which was no doubt reinforced by Communist ideology and American anti-imperialism), the former European imperial powers came to terms with world realities. In any case, Europe's former enemy, the Soviet Union, could not be defeated militarily, and this also applied to the United States. The different experiences of the Cold War, and the different visions of how it was to be ended in Europe and the US, are relevant to the understanding of these structural changes. Geographical proximity meant that the fear of the Red Army and of a Soviet nuclear onslaught was stronger in Europe than in the United States. In this situation, if the Soviet Union attacked Western Europe there always existed uncertainty in the minds of Europeans as to whether the US would come to their rescue at the cost of its own security and even survival. This led to the assumption that Western Europe needed the US more than the latter needed Europe. Throughout the history of the Cold War and after, it has never been easy for Europeans to accept this sense of dependency (in the 1960s de Gaulle demonstrated his resentment against this directly when he withdrew France from NATO's integrated military command), and Europe has become more assertive economically and politically in recent years. Thus, the end of the Cold War liberated Europe from the superpower confrontation, while at the same time, the strategic importance of Europe has declined in America's strategic planning.

Within Western Europe there existed different visions of the end of the Cold War. For instance France wanted a Europe governed by Europeans and led by the French, and free of any single or combined hegemonic control by the Soviet Union, Germany, or the United States. Britain, on the other hand, has consistently pursued a leading role both in Europe and in the world through cooperation with the US, with or without the Cold War. The majority of Western Europeans believed that a slow and almost unremarkable progress with *détente* with the East over the years would be beneficial to their national interests. Thus Western Europeans pursued *détente* with the Soviet Union mainly as an alternative to the end of the Cold War. European *détente* entailed a degree of cooperation within the framework of the Cold War, and this was partially secured by the German Ostpolitik in the 1970s. Whether *détente* delayed the peaceful end of the Cold War or precipitated it is still difficult to answer conclusively. But we now know that the exercise of *détente* in the 1970s helped to soften the rigidity of the state socialist systems in Eastern Europe, while *détente* also encouraged pro-Western Soviet thinkers to promote reforms at home.[34]

For the US, the reduction or even removal of Soviet nuclear weapons (especially the deadly threat of the SS18s) and regime change in the Soviet Union constituted the most important part of the exercise of ending the Cold War. Throughout these years the differences in the strategic culture on both sides of the Atlantic were already becoming apparent: Western Europe emphasised diplomacy, negotiations, economic ties, and gradual evolution, and learned, by trial and error, to live peacefully with Germany, while the US could be more confrontational, coercive, thorough and determined in ensuring that the USSR clearly abandoned its ambitions to compete with, and to defeat, the West. The end of the Cold War meant that Europe had lost its only likely enemy, and it opened up the possibility of moving Europe into the post-modern world whereby the 'emphasis has shifted from the control of territory and armies to the capacity to join international bodies to make international agreements'.[35] This largely left the US, now the sole superpower, on its own, to fend off pressures and conflicts which still exist outside Europe. Meanwhile Western Europeans have begun to integrate the former socialist countries in Eastern Europe into their existing systems, NATO and the European Union, to make Europe for the first time in its history 'free and whole'.

The structural changes during and after the end of the Cold War explain to some extent why the transatlantic alliance does not function as it used to. However, both the US and Western Europe took comfort from the successful end of the Cold War. For the Americans, their hard-line approach paid off, while Europeans would like to think that their soft approach through *détente* must have compelled the East to realise the futility of waging the Cold War. In either case, the West has strong reason to believe that the sacrifices made to wage the lengthy Cold War were worthwhile. In the former socialist countries, there were signs of the decline of authoritarianism, and strong incentives to democratise their societies. Between 1991 and 1999, the UN engaged in numerous major interventions in inter-state conflicts and the rationale behind this was explained by Kofi Annan, the current UN Secretary-General: while the UN charter 'protects the sovereignty of peoples, it was never meant as a licence for governments to trample on human rights and human dignity'. He concluded that sovereignty means 'responsibility, not just power'.[36]

These sentiments have been translated into an expansion of the meaning of the 'threats to international peace and security' in the UN charter and have resulted in the lowering of the threshold for military intervention in other states. The West has become more assertive and interventionist – its values and systems have now become a 'force for change rather than a means of resisting change'.[37] The Clinton administration attacked 'rogue states' (Iran, Iraq, North Korea and Libya) as the West's enemies, who 'refuse to accept and abide by some of the most important norms of practices of the international system', and whose rulers were 'aggressive and defiant', pursued weapons of mass destruction, and remained on 'the wrong side of history'.[38] A similar moral argument, which often justified humanitarian intervention in the 1990s, also comes from Europe. During the Bosnian interventions, Tony Blair stated that:

> We have to enter the new millennium making it known to dictatorships that ethnic cleansing will not be approved. And if we fight, it is not for territorial imperatives but for values. For a new internationalism where the brutal repression of ethnic groups will not be tolerated.

This, according to Robert Cooper, represents a 'classic statement of post modern aspirations'.[39] In the first decade of the post-Cold War years, Europe, the US and the UN redefined the concept of sovereignty, equating state sovereignty with the sovereignty of human beings. Despite the different experiences of the Cold War, and the different security imperatives in the post-Cold War, a consensus emerged on both sides of the Atlantic that the West must take a stand against those countries whose governance was anti-humanitarian and undemocratic.

Human agency

Having seen that the structural changes do not in themselves explain fully the rift in the transatlantic alliance, this chapter will now examine to what extent Bush's leadership style contributed to these transatlantic problems. Bush and his close Republican advisers, sometimes called the 'Vulcans', are deeply suspicious of globalisation, which is, in their view, 'undercutting the authority of individual states'.[40] They regard multilateralism, transnational actors, multinational corporations, and inter-governmental organisations as devices to erode sovereignty and American power. The 'Vulcans' or 'sovereignists' take an almost classical realist view of the world. Instead of targeting a group of terrorists who could get access to WMD, they targeted the Axis of Evil, on the assumption that without the protection and support from such states, international terrorists would not survive. Bush takes power seriously, and believes in 'military might' in dealing with the rogue states and especially Iraq, whose leader could only understand the 'language of military power', not Clinton's 'smiles and scowls' of diplomacy. The Vulcans included neo-conservatives, especially Paul Wolfowitz (the Deputy Defence Secretary until 2004) and Dick Cheney (Vice President), both of whom sold Iraq successfully to the White House as the next target after Afghanistan. After 9/11, Bush himself thought that Clinton's previous containment policy had become 'less and less feasible' and Saddam's capacity to 'create harm' must be punished before harm was done. The idea of pre-emption, enshrined in the 2002 National Strategy Document, was formulated by his National Security Adviser, Condoleezza Rice.

The disintegration of the Soviet Union increased the Vulcans' confidence in America's overwhelming power and influence in the world. The Cold War victory reinforced their belief that America's values were universal, and would continue to prevail. They are not isolationist, but essentially internationalist. The neo-conservatives have their origins in the hawkish wing of the Democratic Party, with Jean Kirkpatrick, who converted to Republicanism and served in Ronald Reagan's inner circle in the early 1980s. The Vulcans put America's national interests first, and are determined to change the status quo (as opposed to the UN's primary role of maintaining it) for the good of Americans, and thereby for the good of the whole world. Iraq was only to be the first step towards the Vulcans' goal of democratising the Middle East in order to remove the fundamentalist Islamist threat to the US. The White House assumed that Iraq with its traditions of leadership 'could become a model for democratisation, which would in turn be beneficial to the US'. John Lewis Gaddis thinks this is 'in every sense . . . grand strategy' – 'what appeared first to be a lack of clarity about who was deterrable and who wasn't, turned out to be a plan for transforming the entire Muslim Middle East'.[41] For others, the Vulcans are identified as optimists. We now know that the Pentagon was not particularly enthusiastic about planning for the future of Iraq after the war. Many of the assumptions the Vulcans had made at the time of the US's invasion have proved wrong. The events that followed after the end of the speedy war in March–April 2003 had not been anticipated. Many more Americans died after the end of the war, with the number of deaths now totalling over 2,000.

It is therefore tempting to ask whether, if the Vulcans had not been in power in September 2001, things might have been different. After all there was no convincing link between Saddam and al Qaeda. Nor was there firm evidence of WMD in Iraq. Just as Vietnam was seen by the John F. Kennedy administration as the key to blocking the onward march of Communism in the early 1960s, Iraq was chosen to eradicate the problems the Americans faced after 9/11. One school of thought argues that the United States has developed the habit of getting into 'unnecessary wars', other than those 'in response to an unprovoked attack on one's territory or citizens or those of a friendly country'. Behind these decisions were 'all the President's men' or a 'war party' determined to push the decision to go to war as the only one open to the US.[42] Prior to the war with Iraq, a former Democrat President, Bill Clinton, voiced his concern about a possible war with Iraq without the full backing of Congress or US allies. He stated, the 'question is not whether to attack Iraq, but how, and under what circumstances'.[43]

There is therefore strong evidence to suggest that the human agency represented by the Vulcans contributed significantly to the Iraq decision. But this must be seen in context. A majority of Americans, including academics and intellectuals, accept that the war in Iraq made the US much less popular with the international community than they had anticipated. They were particularly disturbed by scandals such as the abuse of Iraqi prisoners at the hands of their fellow Americans in Abu Ghraib prison in Baghdad, by mounting anti-American sentiments in the Arab world, by the fragmentation of Iraq, and by what is perceived to be an endless commitment to nation building in Iraq. They still believe in the rightness of the decision to remove Saddam Hussein, but are concerned about the posturing and strong language of the US administration, which often seemed to the international community as signs of America's arrogance. The White House has already spent millions in 'public diplomacy' in an effort to try to improve America's image abroad. The factor of human agency is thus a contributing factor to the acceleration of the Iraq crisis, but did not necessarily determine it.

Conclusion

Through the three levels of analysis (the historical background, the structural factors, and the factor of human agency), it has become clear that the scope and nature of the transatlantic rift over the Iraq war is much more complex than it appeared at the time. The cultural and historical origins of the US suggest that what the Europeans have thought as the ills of the Bush administration – unilateralist, hegemonic, the pursuit of perfect security as protection against the external threat – had all existed long before 2001. It came as no surprise to Americans that the Bush administration reacted to 9/11 by deciding on war against Iraq. Of course, there were those who were against the war without the UN's proper sanction to use force, but, in large measure, the decision to go to war against Iraq was accepted in the US. An examination of the structural factors has explained some of the problems facing the transatlantic alliance, but it also confirms that Europe and the US both became interventionist after the end of the Cold War. Similarly, the factor of human agency appears to have shaped the manner in which the US responded to 9/11, but this does not necessarily explain the substance of its reactions. However, when human agency interacted with the structural changes, the rift seems to have become more discernible, and explainable. The transatlantic rift has been helped by the cumulative structural differences in capabilities, aspirations and strategies between the US and Europe over the last decades. In addition, the cohesion of the alliance has been challenged by a series of major historical events – the end of the Cold War, 9/11, and the war against Iraq. Europe has enhanced its economic and political status in the world as a group of the most developed societies. Almost all Western European countries belong to the list of the top nineteen richest countries in the world. The end of the Cold War has removed the need for Europe's dependence on the US for ultimate security, and has helped to develop a more independent profile in certain areas, such as human rights, environmental issues, and trade and development. The other side of the coin is that Europeans can now afford to become openly divided over such issues as Iraq and can criticise the US. In the case of the Schröder government, his criticism against the US increased the chances of his re-election prospects in the autumn of 2002. The Americans now freely challenge the unity of Europe by dividing it into the new and the old. These moves are all against the rules of the Cold War alliance.

The Iraq crisis also divided European governments into the so-called Euro-Atlanticists and Euro-Gaullists, but both factions could unite in the belief that the EU should be strengthened: in the case of the Euro-Atlanticists, a stronger EU would help to exercise some influence over the US, while for the Euro-Gaullists, this would help to establish Europe as an alternative power centre to the US. The division of Europe over Iraq may have a lingering effect on some bilateral political relations, such as those between Tony Blair's government and Germany (a concern which may have been reduced with the 2005 choice of Angela Merkel as the new German Chancellor to replace Schröder). However, the division has not necessarily affected the fundamental agreement by Europeans on the continuing need for the integration of the whole of Europe. Tony Blair once stated that he 'prefers a divided Europe, part of them in support of the US, rather than a united Europe all against the US', and this explains the diversity of Europeans as well as the strength of the European Union, on which the prosperity of today's Europe depends.[44] Thus, the European Union has not been seriously weakened because of Iraq or because of the divergence over America's policy in Iraq. On the contrary, the Iraq crisis suggests, more than anything else, that Europe has strengthened its hand against the US in the latter's choice of policies. Those Europeans who opposed the war, including Chirac, questioned the wisdom of intervening in Iraq militarily, which might, as it turned out correctly, increase the hostility of Islamic extremists towards the West, thereby making the world an even more dangerous place than before. When President Bush was finally convinced in March 2003 that the French would not join the US in confronting Saddam Hussein, he rightly claimed that 'this wasn't just about Saddam Hussein – it was about the ascendancy of power in Europe'.[45]

The transatlantic rift is therefore also about America's difficulty in carrying the whole of Europe with it, about the limitations of the United States as the sole superpower, about the differing attitudes about the use of force in relation to legitimacy, but not about Europe's inability to resort to military force when needed. This is even more frustrating for the US, since those who favour a renewed and strong transatlantic alliance insist that

America could save political and financial costs if it secured wholehearted support from Europe. This was especially true when the current Bush administration wanted not only regime change but also the democratisation of Iraq and of the whole of the Middle East.

In all fairness, however, those views in favour of a renewed transatlantic alliance are in a minority in the US. The current debate there is about Bush's difficulty in establishing a stable Iraq after Saddam, about the wisdom of the exit strategy from Iraq, and about the reduction of the domestic popularity of the Bush administration. In these debates, the transatlantic alliance figures very little. The same is true for the European debate, which focuses on the reconstruction of Afghanistan, the rising unemployment in EU countries, the impact of the 'war on terror' on the internal security of individual countries, such as Britain after the London bombings on 7 July 2005, the recent crisis over the EU's constitution, and the impact of Turkey's eventual membership of the EU. The strengthening of NATO is more apparent in the form of EU–NATO cooperation whereby the EU is taking on more of NATO's security and defence roles in Europe, such as in Bosnia and Herzegovina, under the UN's mandate. All in all, the transatlantic alliance has been marginalised into becoming something of an obscure institution. The crisis over Iraq has finally moved the alliance into becoming a *bona fide* post-Cold War institution, which has so far failed to establish a clear purpose and sense of mission. This does not, of course, mean the end of transatlantic relations (bilaterally or an *ad hoc* coalition of the willing), as they had existed before the outbreak of the Cold War. Such transatlantic relations can be vital in the age of global anti-Western terrorist campaigns. How long the transatlantic alliance will continue to function as a unified organisation, however, remains to be seen.

Notes

1 The author is grateful to the British Academy for providing a grant to enable her to complete her research for this chapter.

2 See Saki R. Dockrill, 'After September 11: Globalisation of Security beyond the Transatlantic Alliance', *Journal of Transatlantic Studies*, 1:1 (Spring 2003), pp. 1–19.

3 Philip H. Gordon and Jeremy Shapiro, *Allies at War: America, Europe, and the Crisis over Iraq* (New York: McGraw-Hill, 2004), p. 4; Charles A. Kupchan, 'The Travails of Union: The American Experience and its Implications for Europe', *Survival*, 46:4 (Winter 2004–5), p. 115; Marta Dassu and Roberto Menotti, 'Europe and America in the Age of Bush', *Survival*, 47:1 (Spring 2005), p. 110.

4 Thomas S. Kuhn, *The Structure of Scientific Revolutions* (Chicago: University of Chicago Press, 1996), 3rd edn, pp. 43–5, 182–210 ff.

5 Chris Gilligan, 'Constant Crisis/Permanent Process: Diminished Agency and Weak Structures in the Northern Ireland Peace Process', *The Global Review of Ethnopolitics*, 3:1 (September 2003), pp. 23–6.

6 Condoleezza Rice, 'Promoting the National Interest', *Foreign Affairs*, 79:1 (2000), p. 54.
7 'The National Security Strategy of the United States of America', 17 September 2002, p. 27. http://www.whitehouse.gov/nsc/nss.html; Jonathan Monten, 'The Roots of the Bush Doctrine: Power, Nationalism, and Democracy Promotion in US Strategy', *International Security*, 29:4 (Spring 2005), pp. 140–1.
8 Robert S. Litwak, *Rogue States and US Foreign Policy* (Baltimore: Johns Hopkins University Press, 2000), pp. 2–3, 7, 149–50.
9 Ivo H. Daalder and James M. Lindsay, *America Unbound: The Bush Revolution in Foreign Policy* (Washington DC: Brookings Institution Press, 2003), p. 120.
10 Jeremy D. Rosner, 'American Assistance to the Former Soviet States in 1993–1994', in James M. Scott (ed.) *After the End: Making US Foreign Policy in the Post-Cold War World* (Durham, NC: Duke University Press, 1998), p. 242.
11 Gordon and Shapiro, *Allies at War*, pp. 104, 142–4, 152.
12 James Kitfield, 'Fractured Alliances', *National Journal*, 8 March 2003, in NATO ENLARGEMENT DAILY BRIEF (NEDB): http://groups.yahoo.com/group/nedb p. 2, NEDB, 7 March 2003.
13 Gordon and Shapiro, *Allies at War*, pp. 10, 23, 126.
14 Bob Woodward, *Plan of Attack* (New York: Simon & Schuster, 2004), pp. 100–1, 107, 118.
15 BBC News, 3 January 2003, http://news.bbc.co.uk
16 'Divisive Diplomacy with Europe', Editorial, *New York Times*, 11 February 2003, NEDB.
17 The statement comes from Robert Hunter, in 'Iraq Debate Puts NATO to Test', *Washington Times*, 11 February 2003, NEDB.
18 John Keegan, *The Iraq War* (London: Pimlico, 2005), p. 118; Woodward, *Plan of Attack*, p. 294.
19 *New York Times*, 17 March 2003, http://www.nytimes.com; Woodward, *Plan of Attack*, p. 346; Anthony Seldon, *Blair* (London: The Free Press, 2004), pp. 592–4.
20 Gordon and Shapiro, *Allies at War*, p. 76; 'ABC News Poll', *The Washington Post*, 18 March 2003, http://www.washingtonpost.com
21 Gordon and Shapiro, *Allies at War*, p. 3.
22 Mary Nolan, 'Anti-Americanisation in Germany', in Andrew Ross and Kristin Ross (eds) *Anti-Americanism* (New York: New York University Press, 2004), p. 129.
23 Andre Flahaut, Minister of Defence, Belgium, 'An Insult to my Country and its Military', *Wall Street Journal*, Europe, 26 February 2003, NEDB.
24 'McCain, Lieberman, Graham, Bayh Offer Resolution Praising European Allies for Standing with US on Iraq' and 'France Risks Becoming "Irrelevant" with NATO Veto: US Lawmakers', both in 11 February 2003, NEDB; James Mann, *Rise of the Vulcans: The History of Bush's War Cabinet* (New York: Viking, 2004), p. 259.
25 Robert Kagan, *Paradise and Power: America and Europe in the New World Order* (London: Atlantic Books, 2003), p. 3.
26 Kagan, *Paradise and Power*, p. 54.
27 Apart from Kagan, see also Eliot A. Cohen, 'History and the Hyperpower', *Foreign Affairs*, 83:4 (July–August 2004), pp. 49–63; John Lewis Gaddis, 'Grand Strategy in the Second Term' *Foreign Affairs*, 84:1 (January–February 2005), pp. 2–15; Melvyn P. Leffler, '9/11 and American Foreign Policy', *Diplomatic History*, 29:3 (June 2005), pp. 395–413; Jonathan Monten, 'The Roots

of the Bush Doctrine: Power, Nationalism, and Democracy Promotion in US Strategy', *International Security*, 29:4 (Spring 2005), pp. 112–56.

28 Richard Crockatt, *America Embattled: September 11, Anti-Americanism and the Global Order* (London: Routledge, 2003), p. 11.

29 Michael Hunt, *Ideology and US Foreign Policy* (New Haven: Yale University Press, 1987), p. 123.

30 Ibid., p. 133.

31 Gaddis, 'Grand Strategy in the Second Term', p. 52. For the formation of the UN, see Stephen C. Schlesinger, *Act of Creation: The Founding of the United Nations* (Cambridge, MA: Westview, 2003).

32 Philip H. Gordon, 'Trading Places: America and Europe in the Middle East', *Survival* 47:2 (Summer 2005), pp. 94–5.

33 Kagan, *Paradise and Power*, p. 38.

34 Robert D. English, 'The Road(s) Not Taken: Causality and Contingency in Analysis of the Cold War's End', in William C. Wohlforth (ed.) *Cold War End-game* (Pennsylvania: Pennsylvania State University Press, 2003), p. 251. See also S. Savranskaya, 'Transcript of the Proceedings of the Musgrove Conference of the Openness in Russia and Eastern Europe Project, Musgrove Plantation, St Simon's Island, GA, 1–3 May, 1998 (Washington DC: The National Security Archive).

35 Robert Cooper, *The Breaking of Nations: Order and Chaos in the Twenty-First Century* (London: Atlantic Books, 2003), p. 44. For the US's approach to the Cold War, see 'Excerpts from Ronald Reagan's Speech at Hyatt Regency Hotel, Washington DC', Folder: RR Speeches, 1978, in Box 3, Ronald Reagan Subject Collection, Hoover Institution Archives, CA.

36 S. Neil MacFarlane, *Intervention in Contemporary World Politics*, Adelphi paper, 350 (London: IISS, 2002), pp. 50, 52.

37 Lawrence Freedman, *Deterrence* (Cambridge: Polity, 2004), p. 80.

38 Litwak, *Rogue States*, pp. 2, 6.

39 Cooper, *The Breaking of Nations*, pp. 59–60.

40 Daalder and Lindsay, *America Unbound*, p. 42.

41 Peter S. Canellos, 'Despite Cabinet Shuffle, Neocon Ideology Remains', *Boston Globe*, 7 December 2004; John Lewis Gaddis, *Surprise, Security, and the American Experience* (Cambridge, MA: Harvard University Press, 2004), pp. 93–4.

42 John L. Harper, 'Anatomy of a Habit: America's Unnecessary Wars', *Survival*, 47:2 (Summer 2005), pp. 58, 69.

43 'Clinton: Cooperate with Allies', 4 September 2002, http://cnn.com

44 Dassu and Menotti, 'Europe and America in the Age of Bush', p. 107; Gordon, *Allies at War*, p. 131.

45 Woodward, *Plan of Attack*, p. 346.

10

THE SPIES THAT BIND

Intelligence cooperation between the UK, the US, and the EU

Stan A. Taylor

> Liaison is of little interest to spy novelists and is generally ignored in serious scholarly work on intelligence. Nonetheless, this unsung cooperation of foreign intelligence services – in wartime, during the Cold War, or while fighting terrorism; in joint operations, overt support, or sharing of analysis – is responsible for some of the greatest successes of US intelligence. It also has been the source of huge headaches, and not just for our side.[1]

Introduction

Intelligence relationships are different from other political, diplomatic, or even security relationships. As Martin Rudner reminds us, 'they are among the most intimate and enduring' of relationships between nations.[2] Although formal liaison arrangements will always be approved at higher levels, the impetus for such cooperation often comes from lower-level operators or intelligence officers from different countries who share interests, concerns, and information. They may have an urgent need to access information possessed by other services or they may merely share long-run interests. They tend to be less affected by the drift of political and diplomatic relationships that exist above them and there is always the hope that they may garner some tidbit of intelligence that may prevent a major catastrophe.

This chapter makes three assertions. First, the intelligence 'special relationship' between the United Kingdom, the United States, Canada, Australia, and New Zealand (usually referred to as UKUSA) continues as the only formal, multilateral, and (more importantly) significant intelligence sharing arrangement in the contemporary world. Second, efforts to develop a similar relationship between the European Union countries have been largely unsuccessful. The only exception to that statement was the sharing of intelligence about Soviet military capabilities and intentions throughout the Cold War. Third, in spite of the first two assertions, it is the 'spies (a euphemism used to refer to the intelligence professionals and organizations) that bind'.[3] That is, networks of professional intelligence officers operating at all levels of intelligence organizations and reacting to all kinds of foreign policy challenges maintain formal and informal liaisons that mitigate or minimize even official and divergent policies between their states.

This is not a 'conspiracy theory' of any kind. These intelligence organizations and officers do not subvert their state's policies but merely maintain cooperative arrangements against common threats (be they terrorism, weapons proliferation, or anti-democratic governments and forces) regardless of political atmospherics. It is a difficult thesis to prove. Intelligence organizations maintain high degrees of secrecy and intergovernmental intelligence sharing with other services is kept more secret than almost any other aspect of their work.

This chapter begins with a discussion of the relationship between intelligence and security and then turns to a brief history of transatlantic intelligence sharing from World War I up to the present day. It then discusses the impact of the War on Terrorism and the 2003 Iraq War on intelligence cooperation. The chapter closes with a return to the notion that, to some extent, it is 'the spies that bind' relationships between the United Kingdom, the United States, and the European Union much more closely than many believe.

Definitions

The term *intelligence*, as used in this chapter, refers to the collection, analysis, production, and utilization of information collected by one state about potentially hostile states, groups, individuals, or activities. This information is used by national decision makers to enlighten their decisions that affect national security. Intelligence differs from other sources of information in that much, but certainly not all, of the most critical information is collected clandestinely. Intelligence also refers to the government organizations that collect and analyse information as well as to the process by which this function is performed.

The Latin origins of the word are revealing. The Latin prefix *inter* means between or among. Thus, the term *international relations* refers to the relations between nations. The remainder of the word *intelligence* comes from the Latin word *leger* – the gathering of fruit or vegetables. Over time, these two terms were combined into one word referring to the knowledge and skills necessary to make sound horticultural decisions. Gradually, the meaning of the term *intelligence* came to connote the skills and aptitudes needed to make wise and productive choices about any aspect of one's life. It is used in this chapter to refer to the knowledge and information necessary to make informed decisions about statecraft.

Although some think there is a difference between intelligence as used by psychologists and intelligence as used in national security discussions, in reality they come from the same origins. The word intelligence in 'intelligence quotient' or IQ refers to the human ability to process external information into usable and productive knowledge. That is not substantially different from the ability to collect and process information about the interstate system as a necessary prelude to informed statecraft.

The relationship between intelligence, power, and security

Intelligence, in this sense, is ubiquitous. Virtually every state collects, analyses, and utilizes intelligence to one degree or another. The means by which states seek to acquire power may differ according to the geopolitical conditions, political culture, and prevailing ideologies of any given state. But every state, in one way or another, seeks to enhance its security by strengthening its power position relative to other states, or by allying with states who may be better equipped to protect its security.

Two early pioneers in the analysis of state power, Harold and Margaret Sprout of Princeton University, argued that there were five functions or variables by which the power of states could be analysed or measured: (1) The information-providing function (how well do states define a need for information and then collect, analyse, and utilize that information?); (2) The decision-making function (can a nation coordinate all of its resources into an effective strategy?); (3) The means-providing function (how well can states provide the elements of power needed to achieve strategic goals?); (4) The mean-utilizing functions (how well can a government utilize or effectively organize all of the means it may possess?); (5) The resistance-to-demands function (how resilient is a state to the demands and challenges of other states?). This approach to power clearly places intelligence as a first step, perhaps the key step, in the acquisition of state power and, it is hoped, toward greater security.

The theory of intelligence cooperation

Intelligence cooperation between states begins with recognition of common needs. It occurs when the national intelligence agencies of one state share intelligence with other states or enter into informal intelligence liaison agreements with them. It then often results in more or less formal agreements as to collection technology and geographic specialties. No nation has a corner on the market of strategic information, and political exigencies often dictate that states turn to other intelligence services for help. Prior to the latter half of the twentieth century, intelligence cooperation was minimal. Obviously, states would share military information about an enemy with alliance partners during war or conflict, but detailed information about from whom or how the information was collected was seldom, if ever, shared. Normally, when the conflict ended, so also did the intelligence cooperation.

It is not difficult to understand why intelligence sharing has been slow to develop. States have always been reluctant to admit publicly that they engage in secret activities that might violate the privacy, if not the sovereignty, of other states. Intelligence, by its very nature, does that – it clandestinely collects information that other states do not want revealed. In fact, two of the more widely used intelligence collection techniques – human intelligence (HUMINT) and signals intelligence (SIGINT) – involve either clandestinely intercepting various communications to or from (or within) target countries or convincing the nationals of a target state to commit treason. It is a murky field, a 'wilderness of mirrors', whose ubiquity speaks louder than any official documents. In fact, it is somewhat remarkable how little is said about intelligence practices in traditional international law. There is no other way to look at it than to say that states are very reluctant to admit that they spy on one another and even less reluctant to confirm the extent of the intrusive clandestine activities in other states by entering into sharing relationships.

Cross-Atlantic intelligence cooperation

Intelligence sharing across the Atlantic was virtually unknown prior to World War I. Although most European nations had civilian intelligence services by the beginning of the twentieth century, America did not. Excepting military intelligence, no organizations existed in America with whom intelligence could be shared or with whom European services could cooperate. In America, prior to World War I, espionage was seen as something practised by European states still mired in traditional Balance of Power politics and was widely disdained by both the American public and government. Historically, the US had developed intelligence agencies during every war, but had quickly disbanded them when the war was over.

World War I

No nation was less prepared, militarily and in terms of intelligence organizations, than was the US prior to World War I. Perhaps as a consequence, other intelligence services, both friendly and hostile, carried out vast operations in the US. In late 1915, MIC1, Britain's wartime foreign intelligence agency, opened an office in New York City without fully informing the American government of its purposes and subsequent extensive activities. The office was headed by Sir William Wiseman who befriended US President Woodrow Wilson's closest political adviser, Colonel House. House, attracted by Wiseman's accent and baronetcy, met frequently with Wiseman and introduced him to President Wilson. Wiseman became a major source of wartime information to the US government and helped shape US policies during the war. As Andrew has noted, never in the modern history of international affairs has an intelligence agency of one country exercised such a powerful influence on the senior leadership of another country.[4] A much later British ambassador to Washington claimed that Wiseman could easily be 'the most successful "agent of influence" the British ever had'.[5]

At least up to 1917, Germany conducted unprecedented espionage and sabotage in the US in an attempt to keep America out of the war and to prevent American war goods from reaching European nations fighting Germany. In fact, German agents, under both official and non-official cover, were responsible for burning or bombing twenty-two war-related industrial plants (many British-owned) and for sinking or damaging twenty-six US vessels laden with war material and destined for Britain or Russia. In two plant bombings alone, potential war materials worth $24 million were destroyed. All but one of these acts of sabotage, in addition to other covert activities, including acts of psychological warfare meant to influence the American public to stay out of the war, were carried out while America was officially neutral in the war.

Ultimately, American entry into the war was facilitated by British intelligence. Room 40, the location and name of Britain's Naval SIGINT agency, intercepted a telegram from German Foreign Minister Zimmerman that was sent to the German Ambassador in Washington, DC with instructions to be forwarded to the German Ambassador in Mexico. The telegram requested the Ambassador to persuade the Mexican government to join Germany in the alliance against the British. In return for this alliance, Germany, at the conclusion of the war, would help the Mexicans regain much of the southwestern part of US territory. When the British Foreign Office revealed the contents of the Zimmerman telegram to President Wilson, he and the American public changed their attitudes towards neutrality and America entered the war.

After the war, a civilian SIGINT organization was created from a wartime military SIGINT agency that had operated for several years. A former State Department code clerk, William Yardley, headed M-8 (eventually called The Black Chamber) and by the end of World War I, through the 1919 Peace Conference, and at the Washington Naval Disarmament Conference of 1922, this organization gained valuable experience in breaking codes and provided significant information to American foreign policy officials. The Black Chamber, however, was dismantled during the Hoover administration in 1929 by newly appointed Secretary of State Henry Stimson who believed that 'gentlemen do not read each other's mail'.[6]

World War II

America was only slightly better prepared, in terms of intelligence, for World War II than it was for World War I. The Japanese surprise attack on Pearl

Harbor in December 1941 brought America into the war and was called by many the greatest intelligence failure in American history. Multiple investigations of this 'failure' eventually led to the 1947 National Security Act which created America's first peace-time, civilian intelligence service. But even before then, America joined the civilian intelligence game. The wartime Office of Strategic Services (OSS) was created six months after the Japanese attack and was modelled after the British Special Operations Executive (SOE). From then on, to the end of the war and after, transatlantic intelligence collaboration blossomed. The British, with the aid of French intelligence and the Polish Cipher Bureau, broke the German code being transmitted over the highly vaunted Enigma coding machine. The Americans were reading the Japanese 'Purple' code and it was inevitable that cooperation between these allies would increase. US cryptologists were seconded to Bletchley Park (the headquarters of British code breaking), and what would later be called the 'special relationship' began to grow.

The phrase 'special relationship' has many meanings. But they may be classified into at least two archetypes. Some argue that the relationship has extremely broad and cultural depth to it and may even go back to a common historical and cultural heritage between Britain and the US. They believe the relationship has as much to do with culture and language as it does with political and security exigencies. Others believe that it is more specifically related to the common defence needs in the 1940s.[7] Scholars in this latter category, especially those who are grounded in intelligence history, see the more precise role of intelligence needs during World War II as the defining characteristic of this 'special relationship'. In some very real ways, intelligence has been the engine driving the special relationship much more than is often realized.

It obviously has to do with more than merely being allies. The Soviet Union was an ally in the same war but a 'special relationship' did not take root. In fact, '[t]he Soviet Union's most striking intelligence successes during [World War II] were achieved not against its enemies but against its allies in the wartime Grand Alliance: Britain and the United States'.[8] Moreover, to cite merely one of several illustrative examples, both the British and American governments gave multiple warnings to Joseph Stalin that the Germans were planning a major attack against Russian forces in June 1941. Stalin ignored the information, believing that the British and Americans would not help the Russians and were participating in a disinformation campaign to drive a wedge between them and the Germans. This led to one of the worst military defeats in Soviet history. At the end of World War II, transatlantic intelligence sharing, though occasionally strained, had developed into a multifaceted formal and informal set of arrangements. No war in history had been influenced so much by intelligence collaboration, much of which continued after the war.

The Cold War

Cold War intelligence collaboration was almost solely focused on determining the capabilities and intentions of the Soviet Union and its Warsaw Pact allies or on countering the vast Soviet-dominated intelligence services that operated in every Western nation. Aside from the sharing of military intelligence throughout this time, the most significant product of the Cold War was perhaps the formal 'special relationship' between the UK and the US intelligence agencies. The phrase 'special relationship' was first used by Winston Churchill in his famous 1946 Fulton, Missouri talk in which he announced that an 'iron curtain' had descended across Europe. The notion of a special relationship is believed to have originated in the pre-war correspondence between Churchill and Roosevelt. US–British relationships have not always been 'special' and the use of the phrase is often for political ends since both the US and Britain have special relationships with several other countries. But in the field of intelligence, the phrase 'special relationship' does have a specific meaning – it defines a relationship that does not exist with any other nations except those included in two quite rare official treaties.

To formalize what was already an informal arrangement, an agreement between Great Britain and the US was signed in May 1943. It was called the BRUSA (Britain-US) agreement and it increased exchanges of personnel and information between the SIGINT agencies of these two countries as well as Canada, Australia, and New Zealand. Over the next five years, diplomatic and intelligence officials of these five countries refined BRUSA and created a new agreement called the UK-USA Agreement (UKUSA, as it is commonly called). The treaty bound the UK, US, Canada, Australia, and New Zealand to cooperate fully in the collection and sharing of SIGINT information. It even divided collection responsibilities between them based on geographical and other considerations. UKUSA has weathered some difficult times – concern about Soviet penetration of British intelligence, the Suez crisis, and a small episode when New Zealand banned nuclear armed and powered vessels from its harbours in 1985 – but it still exists today as the only acknowledged formal intelligence agreement, voluntarily entered into, between sovereign nations. And over its history, individual UKUSA members, in effect, have broadened the alliance by entering into bilateral intelligence alliances with Germany, Denmark, Norway, Japan, South Korea, New Guinea, Israel, Taiwan, and other countries.

But much Cold War intelligence cooperation occurred beyond the UKUSA SIGINT treaty in *ad hoc* and informal ways. In some extremely significant ways, the Atlantic partners shared vast amounts of intelligence information, aided one another in counterintelligence activities, jointly ran Soviet and Eastern European defectors, and ran joint intelligence operations all geared to, in their perception, enhance their security. It is not insignificant that many times the intelligence officers and organizations collaborated more intensely and shared more critical information than their governments realized. It is important to understand the critical role of intelligence cooperation during the early Cold War years. At least up until 1963, the CIA

> acquired *most* of its clandestine information through liaison arrangements with foreign governments. . . . Liaison provided the Agency with sources and contacts that otherwise would have been denied them. . . . [At times, the] maintenance of liaison became an end in itself, against which independent collection operations were judged. . . . Often, a proposal for an independent operation was rejected because a Station Chief believed that if the operation were exposed, the host government's intelligence service would be offended.[9]

Especially in these early Cold War years, the superstructure of intelligence cooperation was often built on a base created by the intelligence officers on the ground. It was the spies that bound together political relationships and whose efforts often ignored the political winds blowing above them. Additional transatlantic intelligence cooperation was attempted as part of two historical streams of development.

Two streams of European security development

Coterminous with the Cold War was the intelligence cooperation that developed under two separate, but related, security and intelligence historical streams within Europe. The first was the development of a security/defence historical stream which began as the Brussels Treaty and later was symbolized by the North Atlantic Treaty Organization (NATO). The second historical stream was the economic/monetary/political stream flowing towards a single European Union. Full treatments of these developments are readily available and only the intelligence implications are discussed here.[10]

The security/defence/intelligence stream

In spite of considerable efforts to assuage the Soviet Union about Western intentions following the end of World War II, virtually all Western leaders soon realized the imminent danger to the fragile coalition that had defeated the Axis. This realization began at the Potsdam Conference in 1945 but became obvious by 1948 in the face of Soviet attempts to impose its control over countries in Central Europe.

The earliest reaction was an event that bridges back and forth between the security stream and the political union stream. In 1948, with concerns about both the Soviet Union and Germany in their minds, Belgium, France, Luxembourg, the Netherlands, and the United Kingdom signed the Treaty of Brussels, a commitment to the collective defence of Europe. In 1952, in an effort to force German rearmament into a common European defence organization, France proposed the creation of the European Defence Community (EDC). Even though the EDC was eventually defeated by the French Parliament, this led to an alternative way to fold West Germany into a common European military arrangement. The Brussels Treaty nations, along with the US, Canada, West Germany, and Italy, and at a series of meetings in both London and Paris in 1954, created a new international organization called the Western European Union (WEU). The WEU had three main goals: (1) to create a secure Western Europe in which economic recovery could occur; (2) to resist any acts of aggression; and (3) to promote greater unity and integration in Europe. The 1948 Treaty of Brussels also demonstrated the existence of a European commitment to common defence and became an impetus for the US to participate in talks with the Brussels Treaty signatories to find ways to strengthen European defence. These talks, with the additional participation of Canada, turned into the Washington Treaty of 1949 (sometimes called the North Atlantic Treaty). It was this treaty that created the North Atlantic Treaty Organization (NATO), consisting originally of the Brussels Treaty nations plus the US, Canada, Denmark, Iceland, Italy, Norway, and Portugal.

A military structure was later added to this treaty organization and NATO came to symbolize Western commitment first to a policy of deterrence and later to a policy of flexible response in dealing with the Soviet threat. There is some truth to the old cliché that NATO was created 'to keep the Soviets out, the Germans down, and the Americans in [Europe]'.[11] As NATO assumed the 'selective security' functions of the original Brussels Treaty, the WEU fell into some disarray and was nearly moribund for the next thirty years. Even with the addition of other nations, it was almost solely a consultative organization. NATO, on the other hand, became the primary vehicle of Western defence policy.

From the 1950s up until the 1980s, NATO was a major producer and consumer of, primarily, military intelligence about the Soviet Bloc. As the focal organization for any military action against Soviet-led forces, it also became a primary target for Soviet and Eastern European penetration efforts. At least eight Soviet espionage agents or rings compromised NATO security and a former high-ranking Soviet spymaster reported that they knew virtually everything about NATO military facilities and planning. By the mid-1980s, some European federalists sought to revive the WEU and make it the core of a European military component that was independent of the US and NATO. That effort was not successful, but it did prepare the way for later WEU involvement in some minesweeping operations associated with the Iran–Iraq War (1980–8). The WEU even announced in 1984 that it assumed a right to play a role in military concerns beyond its own membership. During the series of crises associated with the break-up of Yugoslavia, the WEU came to play a role alongside NATO forces in support of various UN resolutions.

In regard to NATO itself, the instability ushered in by the end of the Cold War forced NATO to develop a new mission. For the first time in its history it became engaged in actual conflict when it agreed to enforce no-fly zones in Bosnia. It was this conflict that revealed to the European forces in NATO how reliant the alliance was on US intelligence, particularly SIGINT. This information was already being shared with the UK, but this virtually bilateral sharing introduced some tension into the alliance over Britain's commitment to Europe. Was Britain going to continue to rely on UKUSA intelligence, not all of which was shared with other NATO countries, or was it going to wean itself from the US and help Europe develop its own intelligence collection systems?

The economic/monetary/political stream

The process through which the European Union has unfolded and blossomed has also altered the practice of transatlantic intelligence sharing. One could even say the process was born out of an intelligence issue. British and American post-World War II intelligence highlighted a very active international communist movement that had designs on the war-torn countries of Western Europe. These early estimates claimed that an economically devastated and fragmented Europe would be a breeding ground for pro-Soviet political movements. Thus, the European Recovery Program (ERP, often referred to as the Marshall Plan) began in 1947 and, aside from helping put Europe back on its feet, led to the Schuman Plan, Euratom, the Coal and Iron Community, the Common Market in 1959, and, ultimately, the European Union.

The need for special liaison between the NATO stream and the EU stream ended when Britain joined the EU in 1973. Of special interest to the matter of intelligence cooperation, however, are the following: (1) The Treaty of Maastricht, the 1991 treaty that began the development of a common foreign and security policy (CFSP). (2) The 1998 meeting between French and British leaders in St Malo, France where an official declaration called for the EU to have 'a capacity for analysis of situations, sources of intelligence, and a capability for relevant strategic planning'.[12] (3) The 1998 EU Austrian summit where the British reluctantly agreed to move towards a CFSP. (4) The 1999 Cologne Declaration which called for 'maintenance of a sustained defence effort, the implementation of the necessary adaptations and notably the reinforcement of our capabilities in the field of intelligence'.[13] The latter declaration was made while NATO was dropping bombs in Kosovo and, as mentioned above, revealed Europe's near total reliance on

US intelligence. (5) Finally, two meetings (one in Finland in 1999 and the other in Portugal in 2000) called for the creation of an EU military force, to be ready by 2003. Such a force would require the adoption of some sort of common intelligence policy. As Villadsen has said, 'The Persian Gulf and Balkan crises [were] sufficiently traumatic to convey the message that if Europe is serious about achieving the objective of a common foreign, security, and defence policy, there is an urgent requirement for a common European intelligence policy.'[14]

By the end of the first Gulf War, France was leading efforts to create an autonomous European intelligence capability. A former French Defence Minister, Pierre Joxe, noted that, 'Without allied intelligence we were nearly blind. Our extreme dependence on American sources of information . . . was flagrant, particularly in the initial phase. [The US] provided, when and how it wanted, the essential information necessary for the conduct of the [1991] conflict.'[15]

The two streams meet

It is at this point, at least in terms of a common intelligence policy and the more controversial common intelligence institutions such a policy requires, that the two historical streams come together. In relatively minor ways, the WEU bridged both streams throughout the Cold War and even into the post-Cold War period. But two developments merit special attention as the WEU has tried to take 'a leading role in developing and encouraging intelligence cooperation among European states'.[16] The first is the WEU Intelligence Section created at WEU Headquarters in 1995 and the second is the WEU Satellite Centre that became operational in 1997 in Spain. Although the Satellite Centre does not operate its own surveillance systems and only purchases imagery from a variety of different countries (France, India, Spain, Russia, the US, and Canada), at one point it was assumed that the Satellite Centre would later become the core of an EU defence unit as called for in the 1999 Cologne Declaration.

Nevertheless, all of these developments that appeared to be leading to a common European Union Intelligence Service (EUIS) throughout the 1990s have faltered. One observer, citing a WEU audit on defence capabilities, has noted that, at least by 1999, 'there is, as yet, no satisfactory sharing of strategic intelligence, either at the national or international level, that would enable joint European military staff to conduct in-depth analysis of a crisis situation'.[17] In spite of the almost inexorable push for greater and more substantive formal intelligence cooperation and for the institutions necessary to support such cooperation, it is still surprising that so little has occurred. Both within the WEU/NATO historical stream, as well as within the dramatic, almost tectonic shift towards a single European Community, formal and significant bilateral and multilateral intelligence cooperation is minimal, again with the exception of military intelligence during the Cold War. What cooperation does occur is much more informal and much more bilateral than one might suspect. What explains this? Why, in spite of concerted efforts by some Europeans, is most intelligence cooperation and liaison informal and bilateral? Five factors ought to be considered.

First, secret agencies can be very powerful tools of statecraft. As such, states are extremely reluctant to yield any control over them to supranational entities. Along with the power to tax and the power to raise armies, the creation and use of secret intelligence agencies is one of the touchstones of national sovereignty.

Second, HUMINT always begins somewhere. Someone, usually at considerable risk, decides to pass classified information to someone else and the personal safety of the initiator is often carried with the secret. As a rule, the more people who know the source of the information, the greater the likelihood that source will be discovered and punished. And the further that information gets from its national base, the greater the likelihood of discovery. Cross-national intelligence sharing significantly increases the risks of compromise and ultimately diminishes sources.

Third, a nation that enjoys a privileged intelligence relationship with other nations is reluctant to dilute its benefits by sharing that fruit with additional states. Britain is the classic example of this. France has actively tried to break up the privileged relationship between the UK and the US, but with little success. The French Hellios satellite system, developed with Spain and Italy, is still a long way from full spectrum imagery. Absent German financial cooperation, Hellios has helped but cannot touch the information Britain receives through UKUSA.

Fourth, intelligence systems function best with unitary political systems. The French and Dutch blow, in June 2005, to the developing European constitution will not end a half-century of progress towards political and economic union in Europe. But it does not bode well for further formal intelligence cooperation nor, more specifically, for the centralized organizations on which such cooperation must rest. Intelligence organizations are not comfortable with rotating chairmanships and large and diffuse executive authority.

Fifth, the purpose of intelligence, in terms of both structure and information, is to inform statecraft – it is to help foreign policy decision makers select policies that will enhance state security. Shared intelligence assumes that foreign policy goals are sufficiently similar that intelligence will serve the ends of all states involved. That is less and less the case in Europe. As the EU has increased in membership it has also somewhat diluted common foreign policy interests. Moreover, transatlantic differences over the respective role of EU military forces versus NATO, ways to combat terrorism, as well as differing approaches to dealing with Iraq, Iran, and the Israeli-Palestinian issues have created a wedge in diplomatic relations. What happens when friends and allies, states who already conduct both formal and informal intelligence cooperation, find themselves pursuing different policy objectives is illustrated in the case of the 2003 Iraq War.

9/11, the War on Terror, and the Iraq War

The 9/11 attack on the World Trade Center and the Pentagon was a watershed event in America. President Bush, with the support of Congress, declared a War on Terrorism and later invaded Afghanistan, the home of al Qaeda. Operating under the newly proclaimed strategic doctrine of preemption, the US government decided to invade Iraq as well. Iraq had emerged from the 1991 Gulf War with its government intact but under numerous UN sanctions and was forced to accept international inspections carried out by the United Nations Special Commission (UNSCOM). Early in the inspection regime, UNSCOM became aware of an Iraqi government document ordering WMD activities to be hidden from UNSCOM inspectors.[18] This, and other similar practices, convinced UNSCOM that Iraq was attempting to pursue a policy of either removing or hiding certain WMD while at the same time maintaining a sufficient research and production base that they could restart all of the banned activities once the inspection regime was over and the sanctions were lifted. During 1991 and 1992, Iraq dismantled or hid parts of its WMD programs in the hope of quickly ending the sanctions.

The 1995 defection of Saddam Hussein's son-in-law to Jordan prompted Iraq to turn over thousands of hitherto hidden documents to UNSCOM. These documents confirmed what most Western intelligence agencies, as well as the UN inspectors, had suspected, namely that Saddam had preserved both production as well as research and development capabilities, even though he had destroyed some of his WMD stockpile. Moreover, Scud missile production facilities were discovered by UNSCOM and a biological weapons plant that was still producing agents was confirmed by the Iraqi government. Both of these activities had previously been denied.

By 1996, Saddam apparently decided that the inspection regime was lasting longer than he had originally believed it would last and that his policy of superficial disarmament, while maintaining production facilities, was being used to justify the maintenance of inspections and sanctions. This led him to actually dismantle substantive WMD programs. But rather than admit this to UN inspectors, he attempted to keep his compliance with UN requirements secret, most likely for domestic political reasons. He even bragged to both domestic and international friends that he was maintaining his WMD programs while he was actually dismantling them.

In 1997, Saddam came to believe that he no longer had to cooperate with the UN inspection regime. International pressures were building to end the sanctions program and these pressures coincided with the US oil-for-food program. At the end of 1998, after facing considerable harassment, UN inspectors were removed from Iraq and Western intelligence services lost their most reliable source of information about WMD in Iraq. From 1999 on, most intelligence information about Iraq WMD came from defectors, Saddam's opponents-in-exile, or from the intelligence services of other Middle Eastern nations. Defectors are often unreliable, political opponents have their own agendas, and the services of other Middle Eastern nations were often victims of Iraqi denial and deception practices or, in some cases, pursuing their own strategic interests.

The stage was now set for the very influential British and American intelligence estimates on Iraq. These estimates arose, however, out of a context that needs to be understood:

1 In a bizarre turn of events, Saddam was now secretly dismantling some of Iraq's WMD programs while at the same time bragging about his WMD to domestic political supporters and to international friends.
2 The seven years of UN inspections had destroyed some WMD but UNSCOM did not have confidence that it had brought Iraq into UN compliance. In fact, UNSCOM reported that of its approximately 250 'surprise' inspections of Iraqi facilities, 244 facilities had been forewarned by Iraqi intelligence agents who had excellent sources within the UNSCOM support team and only six inspections had actually caught the facilities by surprise.
3 All Western intelligence agencies had underestimated Iraq's weapons development prior to the 1991 Gulf War. Evidence emerged after the war that Iraq was closer to acquiring nuclear weapons than Western intelligence agencies had believed it was before the war. Having been once burned, intelligence services were now overestimating Iraqi weapons and were relying on sources of information that might not have been trusted had the UN inspectors remained in Iraq.
4 Every major intelligence service (American, British, German, French, Israeli, Russian, and Chinese), along with the ousted UN inspectors, believed that Iraq was developing WMD at a more rapid rate than they had previously thought. They believed that Iraq was within five to seven years of having deliverable nuclear weapons. As Pollack puts it, '*no one* doubted that Iraq had weapons of mass destruction'.[19]
5 Without hard information from UN inspectors and facing ambiguous evidence from a variety of sources, intelligence analysts fell back on prevailing presuppositions and assumptions.

With this context in mind, both British and American intelligence official estimates asserted the following: Iraq had continued WMD programs in defiance of UN resolutions, since 1998 (when UN inspections had ended), and Iraq had energized its missile programs, was again working on its nuclear program, had renewed the production and stockpiling of chemical weapons, and was making the greatest progress in biological weapons. These conclusions led to the 2003 invasion of Iraq.

After the invasion of Iraq and following several extended attempts to locate WMD in Iraq, it appears all of these assertions were wrong, except two: Saddam was working to acquire missiles with ranges that exceeded the UN-allowed range and he was continuing to develop biological weapons. It also appears that Saddam had tried to conceal these efforts from UN inspectors. While several analyses of why so many intelligence agencies were wrong have been written, much less has been written about the impact of the invasion on transatlantic intelligence cooperation, especially cooperation in the War on Terrorism.

Intelligence cooperation in the War on Terrorism

Intelligence cooperation during the War on Terrorism has to be divided into two stages. During the first phase, up until the war in Afghanistan, political, military, and intelligence cooperation was at an all time high. France was particularly useful because of its long-established fight against Islamist terrorism within its borders. Often seen as the most vocal European opponent of US policies, France provided very valuable military and intelligence support in this first stage. Britain, America's most supportive European ally, did the same. Near the end of the Afghan war, US unilateralism, coupled with a failure to collaborate with European allies during the war, altered the situation.

The second stage began with the planning and execution of the 2003 invasion of Iraq. This stage was marked by growing diplomatic and political tensions between the US–British alliance and, especially, France and Germany. The precursors of this stage appeared near the end of the war in Afghanistan, but erupted quite dramatically with the invasion of Iraq when relations with France, Germany, and other usually supportive European allies significantly deteriorated. France and Germany were the most vocal European states to criticize American and British policies in Iraq, but widespread dissatisfaction created one of the widest schisms ever seen in the transatlantic alliance. It has prompted many scholars to pronounce the alliance, if not dead, at least moribund.[20] Arguments ranged from assertions that basic and fundamental values between America and Europe were now so divergent that the alliance could not survive, to arguments that Europe was now more 'civilized' than America in its view of the use of force. Some of these arguments assumed that Europe would continue what seemed like its inexorable move towards a single sovereign EU. That assumption can now be questioned.

The French and Dutch 'no' votes on the European constitution in Spring 2005 have created a new environment in which the EU can no longer be seen as gradually, but inexorably, moving towards increased powers for the EU institutions. While it remains to be seen just how the EU will deal with this setback, it is quite clear that certain functions and activities that have been talked about and planned for some time will be deferred. One of those functions will relate to intelligence. It now seems doubtful that the EU will formalize intelligence-sharing functions by creating a common intelligence agency. Such a move would evoke some of the worst fears about elitist Euro-politicians derogating one of the most sensitive of state powers. But it is less clear that this will alter the ongoing practice of informal and bilateral intelligence cooperation and sharing that has developed.

The spies that bind

The absence of formal intelligence-sharing agreements does not mean that intelligence services of various countries do not cooperate, share information, and even run joint operations. This cooperation is often unknown or not talked about, but it creates a dynamic fostered by professional intelligence officers who cooperate with other services more than their governments cooperate. By one estimate, in 2001 the CIA had intelligence relationships with 400 foreign intelligence and security organizations and the Russian FSB had similar relationships with 89 different countries. Most of these agreements were very informal and were sealed either by handshakes or by memoranda of understanding. In 2002 a senior CIA official said, 'alongside military and diplomatic coalitions, [we have] a global coalition of intelligence services. From around the world, from our allies and our partners, we receive and share information. We plan operations together and together in many instances we take terrorists off the streets'.[21] For example, in 2003, intelligence sharing between Russia, the UK, and the US led to the arrest of a man trying to smuggle a heat-seeking missile into the US for the express purpose of firing it at a commercial airliner.

Another less well-known example illustrates this same principle. The previously highly secret Alliance Base, a joint French–American intelligence operation, began as a joint CIA–DGSE operation in 2002.[22] Though largely funded by the CIA's Counterterrorist Centre, it now has intelligence officers from Britain, Germany, Canada, and Australia, in addition to French and American personnel. As Priest writes,

> Alliance Base demonstrates how most counterterrorism operations actually take place: through secretive alliances between the CIA and other countries' intelligence services. This is not the work of large army formations, or even small special forces teams, but of handfuls of US intelligence case officers working with handfuls of foreign operatives, often in tentative arrangements.[23]

John McLaughlin, a thirty-two-year veteran of the CIA and a one-time acting director, reported that this relationship between French and American intelligence officers is 'one of the best in the world'. Apparently, much of the willingness to cooperate began in 1994 when a group of Algerian terrorists captured an airplane and planned to crash it into the Eiffel Tower. The plan failed but, within two days of the attack, French President Jacques Chirac ordered his intelligence agencies to cooperate with the US 'as if they were your own service'.

Even after the invasion of Iraq, when French–American/British relations soured, French and American intelligence officials continued their co-operation and sharing. According to the director of the French domestic intelligence agency, the Directorate of Territorial Surveillance, 'the co-operation between my service and the American service is candid, loyal and certainly effective'. The top French anti-terrorism magistrate, Judge Jean-Louis Bruguiere, claimed that 'the relations between intelligence services in the United States and France [have] been good, even during the transatlantic dispute over Iraq'. Even the French Interior Minister, Nicolas Sarkozy, a pro-EU politician actively seeking to succeed Chirac, said in a Parliamentary debate that 'cooperation between the services of the great democracies [was] perfect'. The fact that a multilateral intelligence group, started between US and French services, could weather some of the worst French–American diplomatic relations in recent years underscores the notion that intelligence cooperation is so important to national security that it is, to some degree, immune from the effects of deteriorating diplomatic and political relations between nations.

Such a need is recognized on both sides of the Atlantic. The American Undersecretary of Defence for Intelligence, Stephen Cambone, instructed all military intelligence agencies to be more open in sharing intelligence with allies. Keeping intelligence from allies, he said, has 'hindered America's ability to work with coalition partners'.[24] Perhaps Undersecretary Cambone also has come to recognize the validity of the biblical injunction, 'Lean not to thine own understanding' (Proverbs 3:5, King James version), or, as the same verse is translated in the New American version, 'On your own intelligence rely not'.

The two terrorist attacks on the London transportation system in July 2005 also illustrate the point. British–French relations, seldom rosy, were on a downward trend. French President Chirac was inadvertently over-heard publicly criticizing some aspects of English culture. Even though Paris had fully expected to be selected as the site of the 2012 Olympic Games, the International Olympics Committee selected London. All of this occurred during the aftermath of the French and Dutch 'no' votes on the EU constitution and amid vaguely muted British gloating as it appeared that Tony Blair's vision of the future EU was ascendant. Then, on 7 July and 21 July, al Qaeda- or Islamist-motivated terrorists launched two attacks in London – the first succeeded, killing over fifty people, wounding over 700, and disrupting transportation, business, and tourism. In the second attack, the bombs failed to detonate but the brazenness of the attack, so soon after the 7 July attack, took an additional toll on business and tourism. However, within three days of the first attack, Scotland Yard and MI5 convened an unprecedented 'intelligence summit' involving senior intelligence officials from the US and over twenty European countries.[25] Participants reported that information was shared freely and all were anxious to help British investigators catch those responsible for the acts. In limited but significant ways, intelligence cooperation falls into the category of functional cooperation between nations characterized by the development of Public International Unions in the late nineteenth and early twentieth centuries. Many of those organizations are now Specialized Agencies of the UN. Indeed, the old theory of Functionalism, although it goes in and out of scholarly popularity, seems to apply in the case of intelligence cooperation.

The intelligence services of countries, particularly in the case of international terrorism, but also in other types of international problems, establish liaison relationships, share intelligence information, and plan and conduct joint operations, even when political/diplomatic relations are in decline. As a former senior CIA official has said, 'the intelligence liaison relationship with a particular country could well be far better than diplomatic relations and may even be the only productive relationship'.[26] One obvious advantage of such liaison relationships is that they do not have to be reported to congressional intelligence committees and they do not require Presidential Directives. In the triangular, transatlantic relationship between Britain, America, and the EU, the 'spies', that is the intelligence services, may not guarantee harmonious diplomatic relations, but one would be foolish to ignore their impact on creating and maintaining ties between these nations.

Notes

1 Charles C. Lathrop (pseud.), *The Literary Spy: The Ultimate Source for Quotations on Espionage and Intelligence* (New Haven: Yale University Press, 2004), p. 242.

2 Martin Rudner, 'Britain Betwixt and Between: UK SIGINT Alliance Strategy's Transatlantic and European Connections', *Intelligence and National Security* 19 (Winter 2004), p. 571.

3 I use this phrase as a title for this chapter for two reasons. First, it describes a phenomenon not usually recognized adequately and, second, it is a play on words drawn from one of the path-breaking books about transatlantic intelligence cooperation: Jeffrey T. Richelson and Desmond Ball, *The Ties That Bind: Intelligence Cooperation Between the UKUS Countries*, 2nd edn (Boston: Unwin Hyman, 1990).

4 Christopher Andrew, *For the Presidents' Eyes Only: Secret Intelligence and the American Presidency from Washington to Bush* (New York: HarperCollins, 1995), p. 39.

5 Cited in Andrew, *For the Presidents' Eyes Only*, p. 39.
6 Although there is some debate as to whether or not Stimson actually used this phrase it does express his feelings and he did use it later to justify what he had done. See David Kahn, *The Codebreakers: A Comprehensive History of Secret Communication from Ancient Times to the Internet*, rev. edn (New York: Scribner, 1996), p. 360. Ironically, Stimson, as Roosevelt's Secretary of War, encouraged the development of SIGINT. What had changed? According to Stimson, America was then at war and it was acceptable to read others' mail.
7 John Baylis, in *Anglo-American Defence Relations—1939–1984* 2nd edn (London: Macmillan, 1984), summarizes these approaches in his preface.
8 Christopher Andrew and Vasili Mitrokhin, *The Sword and the Shield: The Mitrokhin Archive* (New York: Basic Books, 1999), cited on p. 243 of Lathrop, *The Literary Spy*.
9 US Congress, Senate, Select Committee to Study Governmental Operations with Respect to Intelligence Activities, 1976, *Supplementary Detailed Staff Reports on Foreign and Military Intelligence*, Book IV, p. 49.
10 See Baylis, *Anglo-American Defence Relations*.
11 Cathal J. Nolan, *The Greenwood Encyclopedia of International Relations* (Westport, CT: Greenwood Publishing, 2002), s.v. NATO.
12 Cited in Ole R. Villadsen, 'Prospects for a European Common Intelligence Policy', *Studies in Intelligence* (Summer 2000, unclassified edition), p. 82.
13 The Cologne Declaration is cited in Villadsen, 'Prospects for a European Common Intelligence Policy,' p. 82.
14 Villadsen, 'Prospects for a European Common Intelligence Policy', p. 83.
15 Cited in Jeffrey T. Richelson, *A Century of Spies: Intelligence in the Twentieth Century* (New York: Oxford University Press, 1995), p. 424.
16 Jon M. Nomikos, 'A European Union Intelligence Service for Confronting Terrorism', *International Journal of Intelligence and Counterintelligence*, 18 (Summer 2005), p. 193.
17 Luke Hill, 'WEU Audit Finds Weak Spots in European Forces', *Defense News*, 6 December 1999, p. 6; cited by Villadsen, 'Prospects for a European Common Intelligence Policy', n. 59.
18 This section is based on the analysis and conclusion in K. Pollack, 'Spies, Lies and Weapons: What Went Wrong', *Atlantic Monthly* (January/February 2004), pp. 79–92, and M. Aid, 'All Glory is Fleeting: Sigint and the Fight against International Terrorism', *Intelligence and National Security* 18 (December 2003), pp. 72–121.
19 Pollack, p. 81 (emphasis added).
20 See, for example, Niall Ferguson, 'The Widening Atlantic', *Atlantic Monthly* (January/February 2005), pp. 40–4; Robert Kagan, *Of Paradise and Power: America and Europe in the New World Order* (New York: Random House, 2003); Robert Cooper, *The Breaking of Nations* (New York: Atlantic Monthly Press, 2003). Oxford historian Timothy Garton Ash takes a different point of view in his *Free World: America, Europe, and the Surprising Future of the West* (New York: Random House, 2004).
21 James Pavitt, speech, Duke University, 11 April 2002, as cited in Lathrop, *The Literary Spy*, p. 248.
22 The information in the following three paragraphs is based on the report by Dana Priest in the *Washington Post*, 3 July 2005, p. A1.
23 Dana Priest, in *Washington Post*, 3 July 2005, p. A1.

24 Peter Spiegel, 'Pentagon Told to be More Open With Allies', *Financial Times*, 2 June 2005, p. 18.
25 Eliane Sciolino and Don Van Natta, Jr., 'With No Leads, British Consult Allies on Blast', *New York Times*, 11 July 2005, p. A1.
26 Lathrop, *The Literary Spy*, p. 246.

11

TRANSATLANTIC RELATIONS IN A GLOBAL ECONOMY

Joseph A. McKinney

Transatlantic relations have undergone remarkable changes within the past two decades. Before the end of the Cold War, the United States and Western Europe were bound together in many ways by a common threat from the Soviet Union. Consequently, economic issues in the relationship were generally subordinated to security issues. For about a decade after the Cold War abruptly ended, economic issues rose to the top of the transatlantic agenda. Economic linkages were perceived to be the ties that would keep the United States and the European Union from drifting apart. Much attention was devoted to fostering economic cooperation through the development of intergovernmental initiatives such as the New Transatlantic Agenda in 1995, the Transatlantic Economic Partnership in 1998 and the Positive Economic Agenda roadmap of 2002. Civil society dialogues also were launched in the form of the Transatlantic Business Dialogue, the Transatlantic Consumer Dialogue, the Transatlantic Environmental Dialogue and the Transatlantic Labour Dialogue.

After the terrorist incidents of 11 September 2001 in the United States, and the subsequent wars in Afghanistan and Iraq, security issues again took precedence in the transatlantic relationship. However, in contrast to the Cold War era, during which the United States and Western Europe mostly agreed on the nature of the security threat and how to deal with it, this time disagreement over these issues drove a deep wedge between the United States and major countries of Western Europe. Transatlantic political relations reached perhaps their lowest point of the post-World War II period during 2003, and have not yet been fully restored.

Fortunately, transatlantic economic relations seem to have been little affected by the political strains. Transatlantic trade in goods is rapidly expanding, having increased at about a 12 per cent annual rate during both 2004 and 2005. Services trade between the United States and the European Union is increasing even more rapidly, having grown by almost 15 per cent between the first quarter of 2004 and the first quarter of 2005. The direct investment of American firms in the European Union countries increased in 2003 by more than 30 per cent over the previous year, more than twice the overall rate of increase in United States direct foreign investment, and declined only slightly during 2004. Likewise, European Union firms increased their investment flows to the United States dramatically during both 2003 and 2004. Significantly, despite severe strain in United States–French relations over the Iraq War, French firms invested almost twice as much in the United States in 2003 ($4.2 billion) as the United States did in France ($2.3 billion).[1] In the same year, American firms increased their investments in France by more than 10 per cent.

Breadth and depth of the transatlantic economic relationship

The transatlantic economic relationship is distinctive for both its breadth and its depth. Together the United States and the European Union account for approximately 40 per cent of world output, and for more than one-third of world trade. The United States and the European Union account for about one-fifth of each other's merchandise trade, and for more than one-third of each other's trade in services.

As significant as transatlantic trade is to both the United States and the European Union, transatlantic investment is more significant and implies an even deeper level of economic integration. The European Union is the destination of almost one-half of the United States' foreign direct investment flows. In 2003 such flows amounted to $57 billion, and the stock of United States direct investment in the European Union stood at $725 billion. Over 60 per cent of European Union investment abroad has the United States as its destination, and the European Union accounts for almost three-quarters of all foreign investment in the United States. The European Union is the largest foreign investor in forty-five out of fifty states of the United States, and is second largest in the remaining five. In 2003 direct investment flows from European Union companies to the United States amounted to more than $60 billion, and the stock of such investment in the United States stood at $866 billion. Transatlantic flows of foreign direct investment increased fivefold during the 1992–2002 period.

One result of this level of transatlantic direct foreign investment is that the combined sales of foreign affiliates of United States firms in the European Union market, and of foreign affiliates of European Union firms in the United States market, are five times greater than transatlantic trade flows. Sales of United States foreign affiliates in the European Union are more than double the sales of American firm affiliates in the entire Asia/Pacific region, indicating the greater integration of economic activity in the transatlantic region. More than half of the research and development conducted by United States firms outside the country is conducted in the European Union, and transatlantic internet bandwidth is more than triple that between

North America and the Asia/Pacific region.[2] The breadth and depth of transatlantic economic ties is unparalleled.

Importance of the transatlantic relationship to the world trading system

The transatlantic economic relationship occupies a critically important place in the world trading system. As the two dominant economic entities in the world, the United States and the European Union largely determine how the system evolves. No significant improvement in the world trade regime can take place without the cooperation of the economic giants on either side of the Atlantic. The rapid transformation of the global economy in recent years has made cooperation on world trade and investment matters even more important than it was previously. Throughout much of the post-World War II period capital flows were restricted even as trade barriers were being gradually eliminated. But during the past fifteen years more and more countries have liberalized both long-term and short-term capital flows. As a result, capital moves with relative ease and with lightning rapidity around the globe. Capital owners relentlessly seek out those environments where the policy mix is considered most conducive to economic progress. When national authorities attempt to depart from this policy mix, they can find themselves deprived of the capital that is so essential for economic advancement. Therefore, policy makers find their policy discretion circumscribed.

Until recently, transportation, communication and transactions costs were important impediments to international economic integration. Product markets, and certainly service markets, were predominantly national markets. During the past two decades a technological revolution has dramatically driven down both transportation and communication costs, thereby increasingly integrating previously separated markets. National borders still matter, and will continue to matter, primarily because of imperfect contract enforcement among national entities having separate legal regimes. However, much economic activity is becoming global in nature, and this fact, combined with the international mobility of investment and technology, is creating pressures for international harmonization of both economic policies and regulatory regimes. These pressures can be expected to intensify as more and more economic activity becomes internationally contestable.

During the 1980s and 1990s pressures for policy harmonization also intensified because of profound changes in the international trade regime. Prior to the Uruguay Round of multilateral trade negotiations, which began in the 1980s and concluded in 1995, trade liberalization involved primarily the removal of border restrictions to trade in goods. The Uruguay Round broadened the scope of the world trade regime in a number of ways. For the first time services trade was brought under the disciplines of the multilateral trading system. Trade liberalization of regulated service industries such as banking and telecommunications involves not so much removal of border restrictions as changes in the regulatory regimes of the countries involved. For other services, such as legal or accounting services that involve certification and licensure on the national or subnational levels, trade liberalization implies changes in domestic regulations that often prove to be highly controversial.

While liberalization of services trade undoubtedly raises sovereignty concerns, other trade issues introduced in the Uruguay Round are even more controversial. Multilateral agreements on trade-related investment measures (TRIMs) and trade-related intellectual property issues (TRIPs) further circumscribe the latitude of domestic policy makers. While the intention of the TRIMs agreement was to create a more favourable environment for foreign investment by reducing uncertainties, its provisions are sometimes resented as limiting the ability of countries to control and regulate corporations within their territories. Some observers, including influential trade policy experts, have questioned the appropriateness and advisability of bringing intellectual property issues under the rubric of the world trade regime when the World Intellectual Property Organization was already in place to deal with such issues. The TRIPs agreement's provisions for protecting intellectual property have also been attacked on ethical grounds, for in some cases abiding by international standards of intellectual property protections can deprive low-income populations of life-sustaining pharmaceuticals or basic food products.

As the world trade regime has come to address more and more issues, and as economies have become increasingly integrated through both trade and capital flows, labour and environmental groups in high-income countries have developed serious concerns about the effects of globalization. Organized labour fears that capital will flee to low-wage countries, causing a deterioration of labour conditions and labour rights. Environmental organizations suspect that multinational corporations will move operations to countries with less stringent environmental policies, causing a 'race to the bottom' that will erode environmental standards in more advanced countries, or will at least inhibit the raising of environmental standards.

How are these concerns about globalization and the nature of the world trade regime to be addressed? In a provocative article in the *Journal of Economic Perspectives*, Dani Rodrik posits a political trilemma for the globalizing world economy.[3] According to Rodrik, in the present world economy countries can choose any two of three options, but cannot possibly choose all three simultaneously. The three options are: international economic integration, nation states with separate national policies, and mass politics (that is, highly participatory political regimes). In Rodrik's words:

> If we want true international economic integration, we have to go either with the nation-state, in which case the domain of national

> politics will have to be significantly restricted, or else with mass politics, in which case we will have to give up the nation-state in favour of global federalism. If we want highly participatory political regimes, we have to choose between the nation-state and international economic integration. If we want to keep the nation-state, we have to choose between mass politics and international economic integration.[4]

While there is some overstatement in Rodrik's position, it is the case that international economic integration can be nowhere near complete as long as each nation is independently setting its own regulations and standards.

To date, the trend in the world economy has been for international economic integration to narrow the scope for domestic policies in individual nation states – that is, the trend has been toward policy harmonization in order to reduce the frictions of international economic integration. But the public outcry against globalization, which has been triggered largely by the intrusion of trade policy into domestic regulatory regimes, raises serious questions about how far and how fast this pattern can continue. Very often domestic economic policies and regulations are rooted in culture, and countries resist forfeiting policy independence as demanded by international economic integration. By empowering interest groups, the internet has greatly increased the effectiveness of this resistance.

Another possibility for solving Rodrik's trilemma would be to turn the clock back by reversing the liberalized capital flows and trade liberalization measures that intrude behind national borders. That would preserve nation-state autonomy in economic policy making, but would mean forfeiture of the economic efficiencies inherent in international economic integration. Thus far, there seems to be little sentiment for turning the clock back. Individual countries that attempt to do so can expect to discover over time that their economies are lagging behind in technological and economic progress. To populations keenly aware of progress being made in other parts of the world, this will be unacceptable, at least in democratic societies.

The third of Rodrik's alternatives involves the choice of international economic integration in combination with highly participatory political regimes, but implies global federalism – 'a world government would take care of a world market'.[5] This solution, however, is simply not on the cards. A surrogate system will have to be devised to shadow global federalism in economic matters without formal political structures. This can only happen through the close cooperation of the two dominant economic players in the world economy, namely, the United States and the European Union.

Transatlantic cooperation in the economic realm is important for a variety of reasons. An integrated international economy is not only highly productive and capable of raising living standards around the globe, but is also conducive to democratization and the protection of human rights.

The information that is freely available in virtually all parts of the world today makes citizens keenly aware of the shortcomings of failed political systems that are unresponsive to popular desires and that fail to protect basic human rights. The sanctions imposed by capital markets on countries that fail to provide economic freedoms to their citizens, or that are characterized by high levels of corruption, provide powerful incentives for national leaders to bring their nations' practices into conformity with international standards in these matters. The market system virtually demanded by global capital markets as a condition for economic progress often brings in its train political freedoms in addition to the requisite economic freedoms.

The United States and the European Union can promote their own economic interests, while at the same time promoting democracy and the cause of human rights, by working together to make the global economic system function smoothly. Strong cooperation by these two predominant economic entities, and their working together to establish institutions that will enhance global cooperation and mutual trust, can provide many of the economic benefits of global federalism without the formal structure.

Transatlantic cooperation within the World Trade Organization

One of the most important avenues for cooperation between the United States and the European Union is through the World Trade Organization (WTO). The purpose of the WTO is to establish a system of rules to guide the economic interactions of nations, and to provide a dispute settlement system for enforcement of those rules. In this way, the public good of unrestricted trade is provided as a benefit available for all nations. The United States and the European Union were largely responsible for the establishment of the World Trade Organization and for its design. Their record of compliance with its rules is less than perfect, however. Both have made frequent use of contingent protection measures, such as antidumping and countervailing duties, in ways that violate at least the spirit of the multilateral trading system. Both have also, on occasion, failed to abide by decisions rendered by WTO dispute settlement panels.

For example, in 1989 the European Union imposed a ban on hormone-treated beef which was challenged before the WTO by the United States and Canada. The dispute settlement panel ruled against the European Union, and the WTO Appellate Body concurred in 1999, but the ban was not removed. Consequently, the United States levied tariffs in retaliation on imports from the European Union. In 2004 the European Union attempted to justify the ban by citing research purported to link hormones in meat to cancer, but the United States rejected the evidence and kept its retaliatory duties in place. The European Union has requested new panels to rule again on the dispute, but in the meantime the ban on US beef remains in place, as do the retaliatory tariffs on European Union goods. A similar result seems likely in the case currently before a WTO panel challenging the European Union's alleged moratorium on the approval of new biotechnology agricultural products.

On the other side of the Atlantic, the United States was slow to change its tax laws concerning foreign sales corporations to make them consistent with WTO rules, and consequently was subjected to progressive retaliation by the European Union. When the legislation was finally changed, the European Union removed its retaliatory duties but requested a new dispute settlement panel to determine whether the new legislation was WTO-consistent. Also, the United States has thus far refused to implement a panel ruling against its Continued Dumping and Subsidy Offset Act (the so-called Byrd Amendment) which channels import duties collected under antidumping and countervailing orders to the companies filing the petitions. Consequently, the European Union and several other countries have been authorized to levy retaliatory tariffs.

A matter with even more important implications for the world trade regime than these trade disputes is the fact that both the United States and the European Union violate the spirit of the WTO agreement with their highly trade-distortive agricultural policies. In 2004, European Union agricultural subsidies amounted to $133 billion, almost one-half of the total for OECD countries, while United States agricultural subsidies stood at $46.5 billion. However, United States farm support increased by 18 per cent in 2004, whereas European Union support has been declining slightly.[6]

In September 2003 the European Union countries agreed to fundamental reforms in some aspects of their agricultural policy. From 2005, a Single Payment Scheme will be implemented so that most direct payments will not be linked to production, but will instead be linked to compliance with environmental standards, sanitary and phytosanitary standards, and animal welfare.[7] This 'de-coupling' of support payments from production in the European Union should result in much less distortion of international trade. Former European Union Agriculture Commissioner Franz Fischler estimated that these reforms will, when fully implemented, have reduced European Union export subsidies by 75 per cent from their 1993 level, and will have reduced the EU's most trade-distortive support program expenditures by 70 per cent.[8]

The European Union is being pressured to reform its Common Agricultural Policy from a number of angles. The May 2004 enlargement of the European Union increased the amount of its agricultural land by almost 40 per cent, and more than doubled its agricultural labour force.[9] The United Kingdom is insisting on agricultural policy reform as a condition of agreeing to a reduction of Britain's rebate of EU funds. Both the European Union and the United States are under intense pressure from less developed countries to liberalize their agricultural markets in the Doha Development Round of multilateral trade negotiations. Doing so would be one of the more effective ways in which the European Union and the United States could contribute toward a more effectively functioning trading system.

Transatlantic economic cooperation beyond the WTO

While enhanced cooperation between the European Union and the United States within the context of the World Trade Organization is critically important to the successful functioning of the world trade regime, the breadth and depth of transatlantic economic integration necessitates a level of policy coordination and rule-making that is not currently possible in the WTO. The WTO has 148 members, 32 of which the United Nations designates as least-developed countries. The economic conditions in these least-developed countries and the trade issues that they face are vastly different from those of the United States and Western Europe. Yet the WTO operates by consensus of its members. Progress on some issues of great importance to the transatlantic economic relationship will not be possible within this framework without undue delay. Other avenues of cooperation will need to be explored.

For certain issues, the Organization for Economic Cooperation and Development (OECD) can serve the purpose. An example of the United States and Western Europe cooperating to improve the global trading system through the OECD is its anti-bribery convention. Through compromise and much discussion the OECD countries managed to reach common ground on an important issue affecting the international trading system. While enforcement of the anti-bribery convention has been far from perfect, the existence of this institutional arrangement is encouraging economic reforms in less developed countries and making for a more efficiently functioning world economy.

A similar process will be needed to reach agreement across the Atlantic concerning the use of economic sanctions. The United States has frequently used sanctions without due regard to their chances for success, and without adequate consideration of their impact on innocent civilians. Disagreements between the United States and the European Union countries over the use of sanctions have complicated transatlantic trade relations. Sanctions are blunt instruments that have little chance for success without widespread participation from the major players in the world economy. While there will inevitably be differences of opinion concerning the use of sanctions, the United States and the European Union should strive to develop a common framework to assess the advisability of economic sanctions, the circumstances under which they should be applied, and the manner in which they should be applied. The principles of the 'just war' doctrine might provide the basis for such a framework. If the United States and the European Union could reach common ground on this issue, sanctions would be more judiciously applied and would have more efficacy when used.

For other transatlantic economic issues that need to be addressed, the OECD is too broad a forum. John Hancock of the WTO and William B.P. Robson of the C.D. Howe Institute have proposed an open transatlantic agreement for these issues based upon plurilateral agreements using conditional most favoured nation treatment.[10] The nontariff trade barrier codes of conduct negotiated during the Tokyo Round were of this type. While most of these were made multilateral agreements in the Uruguay Round so that all members of the WTO now subscribe to them, the Agreement on Trade in Civil Aircraft and the Agreement on Government Procurement are still plurilateral in character. Transatlantic agreements such as these on issues not within the purview of the WTO, or involving issues where the WTO considers bilateral agreements appropriate, could be consistent with both the spirit and the letter of the multilateral trade regime.

Hancock and Robson suggest that such plurilateral agreements would be feasible in the areas of regulatory coordination, investment rules, competition policy, government procurement, movement of people and trade facilitation.[11] Since trade facilitation has now been included on the agenda of the Doha Development Round of multilateral negotiations, it should first be addressed in the WTO, with the European Union and the United States taking the lead in proposing initiatives. Even on trade facilitation issues, however, the United States and the European Union should be able to attain a deeper level of cooperation on customs regulations and border security issues than will be possible in the WTO at large.

The most contentious trade disputes arise out of differences in regulatory practices. Transatlantic coordination of these practices is sorely needed both to reduce strains in the relationship and to increase economic efficiency. A recent study by the OECD Economics Department estimated that reduction of regulatory barriers to trade between the United States and the European Union would increase United States exports by 17.5 per cent and European Union exports by 23.0 per cent. They further estimated that, as a result of these regulatory reforms, GDP per capita would increase by 1.7 per cent in the United States and by 2.8 per cent in the European Union.[12] These were estimates of the static effects of regulatory reform. Positive dynamic effects on long-run growth rates resulting from the reforms would almost certainly outweigh the static effects.

Mutual recognition agreements covering some products standards have already facilitated transatlantic trade. However, greater transparency in the development of new regulations and more opportunity for foreign firm input on draft regulations and standards are needed. In the sensitive areas of food safety and environmental protection, independent assessments by panels of eminent and respected scientists from both sides of the Atlantic may be necessary in order to instil the public confidence necessary for policy harmonization. The formation of transatlantic panels to work on these issues would foster development of epistemic communities of experts that over time could narrow the divergent perceptions on each side of the Atlantic on such issues as food safety. Transatlantic cooperation could also reduce the impediments to trade that arise out of government procurement practices. The European Union and the United States are the two major signatories of the plurilateral WTO Government Procurement Code. They could further liberalize trade by extending full national treatment to each other's companies for all government procurement contracts and then inviting other countries to participate on the same terms. By transatlantic example, liberalization of government procurement practices within the multilateral system could be encouraged.

Agreement on non-discriminatory investment regulations could be attained in the transatlantic sphere long before such an agreement would be possible in the WTO. Investment regulations are for the most part transparent and stable within the transatlantic realm, but this is not the case in much of the world. Therefore, transatlantic negotiations on investment issues would have a very different character from those conducted under the auspices of the WTO. Whereas multilateral investment negotiations would be concerned primarily with the transparency and stability of investment regimes, transatlantic negotiations would need to focus more on opening up investment opportunities in sectors that have been restricted, and on protecting investors against the adverse effects of arbitrary changes in regulations. A transatlantic investment agreement might well provide the template for a future agreement within the WTO which is presently beyond the realm of possibility in that forum.

Much of the transatlantic foreign direct investment of recent years has been through mergers and acquisitions. Increasingly the competition authorities on both sides of the Atlantic render decisions regarding the permissibility of such activities, and at times they reach conflicting opinions. This adds uncertainty to investment decisions and ultimately lowers economic efficiency. While a great deal of cooperation and information sharing by competition authorities occurs across the Atlantic, if institutional and procedural differences could be narrowed, some degree of transatlantic competition policy harmonization should be attainable long before it is feasible in the multilateral setting.

Transatlantic negotiations on the movement of skilled labour should also be considered. As firms increasingly invest in and supply services across the Atlantic, the need for mobility of skilled labour increases commensurately. Admittedly, in an age of terrorism border security issues make labour mobility difficult. Given the similarities between Western Europe and the United States, however, progress should be possible on this issue through transatlantic negotiations which would not be possible in multilateral negotiations. Steps could be taken to narrow the differences in immigration rules and to make immigration procedures less onerous. Also, mutual recognition of professional certifications would allow for transatlantic employment of engineers, architects, accountants, and other professionals. This would not be easy to attain, and probably would require some narrowing of certification standards to be feasible, but deep transatlantic economic integration virtually demands it.

Changes needed in the institutional framework

Despite the efforts that have been made intergovernmentally and by civil society dialogues during the past ten years with the aim of increasing transatlantic economic cooperation, relatively little has been accomplished. The New Transatlantic Agenda agreed in Madrid in 1995, the Transatlantic Economic Partnership approved in London in 1998, and the Positive Economic Agenda articulated in Brussels in 2002 have been long on generalities and short on specifics. The Senior Level Group of undersecretary-level governmental officials has not been very focused or active in promoting progress on issues of transatlantic concern. The Transatlantic Consumer Dialogue, Transatlantic Labour Dialogue and Transatlantic Environmental Dialogue have not been able to garner much political support for their recommendations. The Transatlantic Business Dialogue has been somewhat more active and can point to some success in the form of transatlantic agreements on mutual recognition of testing and conformity assessments. Even this group, however, has been discouraged by failure to implement its recommendations.

In 1998 the European Commission, under the leadership of Sir Leon Brittan, proposed an initiative called the New Transatlantic Marketplace. It called for deeper integration through removal of technical barriers to trade, removal of all industrial tariffs by 2010, free transatlantic trade in services, and liberalization of investment, government procurement, and intellectual property protection. This proposal was not adopted, largely because of opposition from France arising from its concerns over audiovisual services trade.

More recently, Fred Bergsten of the Institute for International Economics and Caio Koch-Weser, the German Secretary of State for Finance, have proposed the establishment of a 'G-2 caucus' consisting of the European Union and the United States, 'to function as an informal steering committee to both manage their own economic relationship and to provide leadership for the world economy'.[13] Bradford and Linn, noting that a weakness of the G-8 summits is that they do not include the major emerging market economies, recommend that the proposed G-2 be complemented by a G-20 which would include major emerging market economies and would replace the G-8.[14] In such an arrangement, specific transatlantic concerns could be addressed in G-2 meetings, but then discussion would be carried over into G-20 meetings to the extent that the issues had relevance for the world trade regime.

While these changes certainly would enhance transatlantic cooperation, it may be that truly effective cooperation between the United States and the European Union will require the establishment of a more formal and well-funded institution, such as a Transatlantic Economic Commission. Such an institution could lower the costs of governments getting together to negotiate agreements, and could act as a catalyst for agreements. It would make possible economies of scale in decision making since mechanisms established to deal with particular issues would already be in place for handling other issues as they arise. Also, experience gained in dealing with specific issues would enable subsequent issues to be dealt with more effectively. If the mandate of the institution were broad enough, there could be economies of scope as well, for in dealing with several issues at once the possibility would exist for trade-offs and side payments that would facilitate agreement.

If such a commission were funded well enough to employ a first-rate research staff, the objective and reliable information provided would facilitate transatlantic agreement on issues in dispute. When negotiating parties do not have access to the same information, or do not completely trust each other, reaching agreement is difficult. This has definitely been a problem in the transatlantic relationship with regard to food safety and biotechnology issues. Additionally, by estimating the distribution of benefits from proposed measures and making those estimates public, concerns over how benefits would be distributed could be allayed. All of these benefits would be contingent upon having in place a research staff that was respected, that was insulated from political pressures, and that had unquestionable integrity.

Establishment of a high-profile institution, with high-level political endorsement and support, to focus on transatlantic economic issues would send a signal that the transatlantic partners were committed to resolution of the issues important to the functioning of the global economy. Uncertainty concerning future policies would be reduced. This would lower future bargaining costs for the governmental entities involved, and would help to reduce the risk premium from investment decisions, thereby ensuring a more efficient pattern of investment.

Summary and conclusions

The transatlantic economic relationship is the largest and most important bilateral relationship in the world, and the level of economic integration is deeper than anywhere else. Given their dominant presence in the world economy, the European Union and the United States must provide leadership for positive developments in their own economic relationship and in the world trade and investment regime. Throughout much of the post-war period the United States provided this leadership. However, since the enlarged European Union and the United States are of approximately the same size and importance to the world economy, neither can effectively act alone as hegemon. Cooperative leadership is required.

To some extent this leadership can be exercised within the World Trade Organization. Indeed, in many respects the current world trading system is functioning quite well because of transatlantic cooperation on multilateral trade issues. But both the European Union and the United States can further strengthen the world trade regime by less frequent resort to the administered protection measures of antidumping and countervailing duty petitions. They can lend legitimacy to world trade rules by more prompt and complete implementation of decisions of WTO dispute settlement panels. They can work together for a timely and meaningful conclusion to the Doha Development Round of multilateral trade talks by reforming their agricultural support programs to remove the trade-distorting elements from them.

However, for some issues the transatlantic economic relationship cannot wait for further progress within the multilateral arena. Deep economic integration across the Atlantic gives rise to a different set of issues from those faced by the majority of the WTO members. Some of these issues are appropriately addressed within the OECD, such as the anti-bribery convention and the development of ground rules regarding the use of economic sanctions. Others, however, are more specific to the transatlantic relationship and need to be solved through common agreement between the United States and the European Union. Included in this category would be issues such as regulatory coordination, investment rules and competition policies.

The structural and institutional framework established thus far for addressing these deep integration issues is inadequate. Good ideas for reform have been put forward, such as those in the New Transatlantic Marketplace proposal, but not much has been done to implement them. The establishment of a more formal and well-funded institution, such as a Transatlantic Economic Commission, could give impetus to the process and over time more than compensate for its costs through increased economic efficiencies.

In addition to the economic benefits arising from greater transatlantic cooperation on economic matters, political benefits would accrue as well. The United States needs to demonstrate its willingness to seek common ground with the European Union, and working together on common economic interests is a logical place to start. The European Union would have the opportunity to demonstrate through cooperation on economic matters the value that it places on the transatlantic relationship, while at the same time following a somewhat more independent course in other areas. Common economic interests may well be the cords that can bind the transatlantic alliance together. A stronger structural and institutional framework for the economic relationship will be necessary to make the bond as strong as possible.

Notes

1 D. Hamilton and J. Quinlan, 'Executive Summary', *Partners in Prosperity: The Changing Geography of the Transatlantic Economy*, Center for Transatlantic Relations, online. Available at <http://transatlantic.sais-jhu.edu/PDF/publications/Executive_Summary-English.pdf> (accessed 20 June 2005).
2 Ibid.
3 D. Rodrik, 'How far will international economic integration go?' *Journal of Economic Perspectives*, 14, 1 (Winter 2000), pp. 177–86.
4 Ibid., p. 180.
5 Ibid., p. 182.
6 R. Minder, F. Williams and J. Blitz, 'OECD study shows scant reduction in subsidies to farmers over past 10 years', *Financial Times*, 22 June 2005.
7 Commission of the European Communities, *Enlargement and Agriculture: Commission Tables Amendments to Take Account of CAP Reform*, Press Release, 27 October 2003.
8 D. McCoy, 'De-coupling key to reform of CAP', *Belfast News Letter*, 10, 17 April 2004.
9 J. Pelkmans and J. Casey, 'EU enlargement: external economic implications', *Intereconomics* (July/August 2003), pp. 196–207.
10 J. Hancock. and W.B.P. Robson, *Building New Bridges: The Case for Strengthening Transatlantic Economic Ties*, British-North American Committee (Toronto: Ricoh Canada Inc., May 2003), p. 20.
11 Ibid., pp. 21–3.
12 Organization for Economic Cooperation and Development, 'The benefits of liberalising product markets and reducing barriers to international trade and investment: the case of the United States and the European Union', *OECD Economics Department Working Papers*, 432 (June 2005), pp. 27–8.
13 C.F. Bergsten, and C. Koch-Weser, 'The G-2: a new conceptual basis and operating modality for transatlantic economic relations', in W. Weidenfeld *et al.*, *From Alliance to Coalitions – The Future of Transatlantic Relations* (Gutersloh: Bertelsmann Foundation Publishers, 2004), p. 248.
14 C.I. Bradford and J.F. Linn, 'Global economic governance at a crossroads: replacing the G-7 with the G-20', *Policy Brief #131* (Washington, DC: The Brookings Institution, April 2004).

12

POLITICAL ECONOMY

Divergence and convergence between the United States and Europe[1]

Craig Phelan

The recent emphasis on contrasting foreign policy orientations in Europe and the US, so forcefully presented by Robert Kagan, has its counterpart in the realm of domestic social and economic policy. Indeed, Kagan's dualistic paradigm ('Europeans are from Venus, Americans are from Mars') has its most ardent adherents not among diplomatic scholars but among students of political economy. For it is when discussing such weighty matters as the welfare state, labour policy, and general orientation to the market economy that Europeans are most comfortable promoting their collective identity and their distance from American values. While a foreign policy that envisions a Kantian 'perpetual peace' rather than the 'anarchic Hobbesian world'[2] is a relatively recent development that could arguably be based on making a virtue of necessity, it is in the sphere of political economy that many Europeans have long been convinced that not only is Europe fundamentally different from the US but vastly superior to it.

Proponents of Social Europe often employ Kagan's dualistic paradigm. Timothy Garton Ash, although often critical of Kagan, suggests that 'a European *identity* can and should be built upon these differences – or superiorities'.[3] Will Hutton argues that 'Europe's welfare states, trade unions, labour market regulations and belief in the husbanded or stakeholder enterprise – along with the role played by government – are not economic and social aberrations. . . . They define Europeanness. They are non-negotiable European realities.'[4] When protesting the war in Iraq in early 2003, two venerable European philosophers, Jürgen Habermas and Jacques Derrida, seeking to illuminate the pre-eminence of Europe over America, focused on political economy as well. In contrast to America, Europeans could take pride in their 'trust in the civilizing power of the state, and their expectations for its capacity to correct market failures'. Europeans were also much more likely to insist on 'social justice' and resist 'the individualistic ethos of market justice that accepts glaring social inequalities'.[5]

According to this popular argument, America's excessive reliance on the market economy, stunted welfare provisions, inadequate public services, stigmatisation of those in need of public assistance, flexible labour laws, and weak levels of unionisation represent the converse of 'Europeanness' and, for some, barbarity itself. 'Most civilized nations compensate for the inadequacy of wages by providing relatively generous public services such as health insurance, free or subsidized child care, subsidized housing and effective public transport', writes Barbara Ehrenreich. 'But the United States, for all its wealth, leaves its citizens to fend for themselves – facing market based rents on their wages alone. For millions of Americans, that $10 – or even $8 or $6 – an hour is all there is.'[6] A flexible labour market and weak social provision may be effective approaches to creating jobs and establishing sound public finance, but they have also produced the most unequal advanced industrial society, with an uncomfortable percentage of its citizens trapped in low-wage service jobs, without adequate health care, adequate pensions, or access to skills training or child care. Rather than a welfare state, America imprisons the socially and economically excluded, which has resulted in world-leading incarceration rates. With millions in prison and millions more under forms of penal control outside of prison, the US has been described by some as more 'a "penal state" rather than a "welfare state"'.[7] The noted author John Gray concluded that 'No European country – not even the United Kingdom – is ready to tolerate the levels of social dereliction produced by the free market in the United States.'[8]

The race to the bottom

The economic side of the Kagan paradigm is far gloomier and more deterministic than the foreign policy dimension. In matters of diplomacy, it is plausible to suggest that a more soft-handed European effort is an eminently viable and welcome counterbalance to a militaristic America. But in the economic realm, we are told, globalisation is already busily at work undermining 'non-negotiable European realities'. Without an all-out effort to create and sustain a powerful European economic counterbalance, globalisation will inexorably recreate across the European continent (and the entire world) the brutal economic landscape of America itself. The need for such an economic counterbalance is one of the most compelling arguments in favour of greater political integration in the European Union, the one real hope of preserving and expanding Social Europe.

In much of the popular literature on the subject, and in many of the pronouncements of governments of both the left and the right, globalisation is portrayed as a strait-jacket of neo-liberal logic from which not even Houdini could escape. According to this view, globalisation is a powerful threat to such standard features of European political economy as neo-Keynesian monetary policy, strong state interventionism, high tax rates, and generous welfare state expenditures. Globalisation requires increasingly low debts, deficits and inflation rates (which are the bedrock of monetarism), privatisation and deregulation, the liberalisation of labour markets, and the scaling back of welfare programmes. Whether left, right or centre, governments now have virtually no choice but to follow the new global logic, so the story goes.

'As markets become more and more integrated and international, and restrictions and controls on the cross-boundary flow of goods, services and capital are rapidly eliminated', opines Robert Went, 'wages, working conditions, employment and social security risk being sucked into a downwards spiral.'[9] In a truly free trading global market, advantage lies with firms whose costs are low, whether these are labour costs, taxes, or regulatory costs. It is this constant pressure to reduce wages and costs that creates this worldwide 'downwards spiral' or 'race to the bottom'. Europeans have already discovered to their dismay that maintaining tightly regulated labour markets and expansive welfare states makes them uncompetitive globally. In his well-received *False Dawn*, John Gray argues that 'In the long haul of history, Europe's social markets may be as productive as American free markets. In the short run, in terms of rivalry in a global free market, they simply cannot be cost-competitive.'[10] Michael Lind was even more downcast: 'It is difficult to resist the conclusion that civilized social market capitalism and unrestricted global free trade are inherently incompatible.'[11]

One aspect of this 'race to the bottom' scenario is worth highlighting: the idea that global market forces have dramatically undermined state autonomy in national decision-making. Forced to follow the dictates of the global market by liberalising labour laws, reducing taxes, cutting budgets, and at least curtailing the continued expansion of the welfare state, globalisation has severely limited the nation state's ability to manoeuvre and its control over the economic forces and actors within its borders. Globalisation has decimated the 'authority, legitimacy, policymaking capacity, and policy-implementing effectiveness of the state'.[12] Since both left- and right-wing governments are forced to cut budgets and liberalise, globalisation has undermined choice in the political spectrum, collapsing differences between the left and right.

The only effective check on the inexorable 'race to the bottom' is the expansion and continued integration of the European Union, which would provide a social market big enough to sustain economic growth and large social outlays. In short, European integration is the best way to prevent European Venusians from becoming American Martians. In the face of the palpable blows to the European project dealt by French and Dutch voters in June 2005, pro-European journalists claimed that, rather than defeat, the rejection of the EU constitution represented an opportunity for Europeans to take stock and recommit themselves to a more social future. Rather than being a tragedy, the votes in France and the Netherlands should revitalise the dream of Social Europe. These referendums will force the EU to more forcefully articulate its 'vision' of 'a new kind of Europe', one that pursues 'harmonisation rather than competition', argues Jonathan Steele. 'Instead of a race to the bottom, Europe should raise social standards to a common level throughout the union, in maximum hours worked, minimum wages, trade union rights and welfare provision.'[13]

The political economy side of the popular paradigm seems to have much to recommend it. It contains a clear internal logic and a compelling narrative. It has an attractive push (the US) and pull (the idea of Social Europe), and its dire warnings about a possible bleak future guarantee that it will capture both headlines and popular imagination. Yet the paradigm ultimately founders on too many generalisations that do not bear scrutiny, and it brushes far too many complexities under the carpet. Its attempt to demonise the US and deify Europe for the sake of promoting Social Europe makes for better populist rhetoric than the basis of sound social and economic policy. And it rests on an ahistorical, snapshot view of events largely coloured by the heightened emotions on both sides of the Atlantic in the wake of the US-led war against Saddam Hussein.

In this chapter I dissect this paradigm by providing a more nuanced view of three of its key tenets. First, I address the question of globalisation and whether or not the pressures of international trade in fact account for the myriad problems that beset European social welfare provisions. And here I conclude that European states face harsh economic decisions that stem largely from internal rather than global causes. Second, I assess whether globalisation has indeed undermined state authority and room for manoeuvre in Europe, and I conclude that while the role of the state is changing in the face of new economic realities, welfare programmes and labour policies are still determined by nation states in traditional ways rather than by any inexorable international economic forces. Third, I challenge the very idea of 'Europeanness' and 'non-negotiable European realities' by assessing the very real differences in social welfare policies and orientation to the market that exist on the large and impressively diverse continent, differences so significant that thoughts of harmonisation along a European model are as yet pipedreams. In the final section I take a longer view of events and place the paradigm in historical perspective, one that suggests that attempts to construct a European identity or social policy must be based on a more solid foundation.

The spectre of globalisation

Although a highly contentious term with a maddening array of meanings, most agree that globalisation is a real and measurable phenomenon. Most notably, globalisation is defined by the explosive growth in the international financial markets and in international trade. The $10 to $20 billion traded every day in the currency markets in the 1970s became $1.5 trillion a day by 1998. Foreign direct investment flows leapt from $60 billion in 1979 to more than $1.2 trillion in 2000. The size, power and influence of multinationals is a dominant feature of globalisation as well, since they are the most strident voices for lower tax rates and better business environments, with the ever-present threat of relocation to back up their demands.

Globalisation is real enough, but excessive fears of its impact have obscured the true nature of the problems facing advanced industrial economies, the continuing authority of the nation state, and the enduring differences between national political economies. In the arena of social welfare policy, for instance, current scholarship clearly indicates that globalisation has very little to do with current crises, and the catastrophic literature that suggests an inevitable convergence toward a minimalist, American-style welfare state is off the mark. For advocates of Social Europe, sound policy formulation must commence with an appreciation of the true nature of the problems rather than a perpetuation of globalisation myths and simplistic dualisms.

Welfare states in Europe and the US, and indeed in all advanced economies, are in transition. All face budgetary pressures that will no doubt only increase in the years to come. Welfare states of all stripes 'have reached a situation of permanent austerity'.[14] Although globalisation is often put forward as the culprit, the problems facing welfare states in advanced industrial economies can be more easily explained by other, less cataclysmic-sounding factors that stem from maturing economies and state programmes. Those factors include: a natural slowdown in productivity that accompanies the shift from manufacturing to service employment; a decades-long process in which welfare states have expanded their commitments to more and more citizens; the demographic reality of an ageing population; and the transformation of households. Understanding these and other factors is necessary in order to analyse how states have responded.

Sustained economic growth is directly related to increasing productivity, and in recent years advanced industrial countries have experienced slower productivity growth. This is a basic economic dilemma facing Europe and the US, yet it has virtually nothing to do with increasing world trade which should, if anything, be a competitive spur to greater efficiency and productivity. Instead, slower productivity growth is a factor of the massive, inexorable shift of employment in advanced industrial economies from manufacturing to the service sector. Service employment is comparatively labour-intensive, less productive and low-wage, which exerts an enormous strain on welfare states through loss of tax revenues and an ever-growing pool of citizens in possible need of assistance. However states respond to this dilemma, there are costs. Liberal advanced industrial nations pursue policies that encourage the expansion of low-wage, private-sector service employment, but the social costs remain high. Poverty levels and inequality tend to increase as a result of these policies, with attendant social problems. More social democratic nations tend to expand the public service sector, which helps to keep a lid on worsening wage inequality but leads quickly to prohibitive budgetary expenses that are difficult to shed as public service employees fight to keep their jobs, salaries and benefits.

A second reason for the current crisis facing welfare states is that, after years of expansion and generous provision, they now consume a share of national wealth large enough to become a matter for public debate. Despite the global economic troubles that began in the 1970s, welfare state expenditures continued to grow substantially as more people were covered, as existing programmes were expanded, and as new programmes were developed. By far the two most expensive items of social provision are pensions and health care, which together account for nearly two-thirds of all social spending. And due to expanded coverage and increasingly generous programmes, the cost of health care and pensions has more than doubled as a percentage of GDP in advanced industrial economies. Between 1960 and 1990, health care and pension commitments rose from 7.1 to 14.3 per cent of GDP in OECD countries. Political and economic problems are natural when spending on social provision grows much faster than the rate of national income. When the tax burden falls on employers, rising costs discourage new hiring and contribute to high unemployment rates. When the tax burden falls on the voting public, political divisions are inevitable and the question of work incentives is bound to arise. Again, these problems are grave but have little to do with globalisation.

A third cause of social welfare distress is demographic, a combination of falling birth rates in OECD countries, longer life spans, and mature social programmes. In OECD countries in 1960, the percentage of the population over the age of 65 was 9.4 per cent. By 2000 it was 13.9 per cent. And when the baby boom generation, who are now in their peak earning years, reach retirement age between 2010 and 2035, the over-65 cohort will increase even more. Put another way, for the OECD as a whole, the ratio of the working-aged (15–64) to those over 65 will fall drastically from 7.5 in 1960 to 2.5 in 2040. The economic consequence of ageing is obvious. Between 2000 and 2030, pensions in OECD countries will gobble up an additional 3.9 per cent of GDP. In addition, the elderly are the biggest consumers of health care, and an ageing population is certain to push up the cost of health care. In OECD countries, ageing alone will account for an additional 1.7 per cent of GDP increase in health care provision by 2030. Like other changes already discussed, the issue of ageing is not a consequence of globalisation.

A final group of problems facing welfare states relates to gender – the dramatic increase in women's labour force participation, and the increase in single-parent households. Although impacting nations in varying degrees, the impact of women working has been profound and complex. In eleven nations where full statistics are available, women's labour market activity jumped from 33.3 per cent in 1960 to 61.9 per cent in 1996. On the one hand, this has been a boon to the fiscal health of the welfare state, as large numbers of women working can mean less reliance on state support and, equally important, increased tax revenue. In the long run women's greater workforce participation places a strain on the welfare state by increasing the need for assistance in social tasks that were previously carried out by women in the home, including child care and care for the elderly and disabled. Publicly-funded care for children, the elderly and the disabled is an expensive proposition, yet these are expanding programmes in most countries. Even where such care is provided privately, there is enormous pressure on governments to subsidise such care through tax credits and transfer payments.[15]

One might question the importance of distinguishing whether the challenges facing welfare states are a product of globalisation or similar sets of largely domestic problems. But the distinction is critical since not all nations face these same problems to the same degree, and therefore one should not anticipate similar policy responses. In 2025, for example, 24.5 per cent of the population of Germany will be over 65, whereas in the UK that figure will be 19 per cent. In 1995 the fertility rate for women in Italy was 1.17, while it was 1.71 in the UK and 1.73 in Sweden. The same national differences can be seen in the extent of shifts toward the service economy, female participation in the labour force, and all other pressures that have generated a sense of crisis in welfare states. 'Distinct countries, distinct problems'[16] remains as true today as it ever was, despite the forebodings of a uniform 'race to the bottom'.

Nation states and national autonomy

The idea that nation states are becoming increasingly helpless in the face of globalisation is important in the intellectual arsenal of both global pessimists and many advocates of Social Europe. The veracity of this idea has generated a great deal of scholarly debate, but several points are clear. First, studies have demonstrated that rather than inevitably leading to budget slashes and tax cuts, trade and financial openness strongly correlates to increased taxation and public spending.[17] Second, since trade openness also correlates to increases rather than decreases in state spending, there remains ample room for states to pursue diverse policy objectives.[18] And third, leading scholars have attacked what they regard as 'the myth of the global corporation' and have demonstrated that states still retain much of their ability to control the activity of multinationals not only within their own borders but worldwide.[19]

To be sure, the pressures of globalisation have certainly limited the effectiveness of some traditionally important measures of state control over the economy. Certain monetary and fiscal policies to fight recession, tariffs and import quotas to protect national industries, nationalisation of industries, and massive subsidies to declining industries are all either more difficult or simply unthinkable in the global environment. But there remain innumerable policy choices available to nation states to control and encourage the economy within their borders. Contrary to the 'race to the bottom' thesis, nation states continue to tax capital with little fear of capital exit, since, surprisingly, there is almost no correlation between corporation tax rates and capital mobility. In both Europe and America, 'governments still have significant – although more restricted – margins of manoeuvre', and 'rather than convergence toward any given set of policies or spending levels, countries have continued to differ widely in what they spend, in how they spend, and in how much they spend'.[20]

The myth of Europeanness

As the above discussion of the continuing importance of national differences suggests, the most glaring deficiency of the popular paradigm is that it rests on the false assertion that there are in fact 'non-negotiable European realities', that there is an inchoate European identity, which can be compared favourably to American realities and an American identity. Just as much of the literature on globalisation tends to flatten national differences, so too does much of the pro-Social Europe literature. And nowhere is this more the case than when discussing political economy.

Rather than the ideal of social democratic uniformity, the European economic reality is one of remarkable national contrasts with no evidence of convergence. An examination of these existing economic differences can gainfully commence with the question of the varieties of capitalism that have developed and persist in the various European countries, a question that has recently commanded the attention of scholars. Even if one ignores the question of which nations actually constitute Europe, even if one ignores the ten new entrants to the EU and focuses on the core Western European nations, there are at least three varieties of capitalism that help explain each nation's unique economic arrangements in regard to the following: the structure of business relations, meaning the nature of interactions between firms and between industry and finance; the system of state relations with both business and labour; and the system of industrial relations, including wage bargaining, labour–management relations and the role of the state in labour regulation. The three models represent ideal types which do not mirror realities in any one nation. And a strong case has been made that each European country has its own distinctive variety of capitalism and that models obscure more than they illuminate. But modelling does help cluster the wide variety of different economic paths that exist in the core states of Western Europe.

The three varieties of European capitalism are (1) state capitalism, in which a strongly interventionist state actively directs economic activity and dictates the behaviour of all economic actors. Until its reforms in recent years, France was the archetypal example of state capitalism in Europe. In (2) managed capitalism, the state encourages economic actors to cooperate and coordinates their agreed upon activities. The Netherlands has been put forward as the best example of managed capitalism, and among larger European states, Germany can be put into this camp as well, even though much industrial policy is shaped by the states rather than the national government. Under (3) market capitalism, a liberal state permits economic actors to conduct their affairs autonomously and make their own decisions about their activities. The US of course has long been depicted as the prime example of market capitalism, and in Europe, post-Thatcher Britain is usually regarded as the closest approximation.

In the last two decades of the twentieth century, in the face of globalisation and other pressures, each European nation has made alterations in its economic relations, and a few scholars have suggested that these shifts represent convergence toward the market capitalist model of the US and Britain. 'The question social market economies face is not whether they can survive with their present institutions and policies – they cannot.'[21] But John Gray's apocalyptic vision has not found many adherents among knowledgeable economists. More popular among academics is the 'varieties of capitalism' school which suggests that state capitalism has disappeared in the face of new economic realties, leaving just two models, market and managed capitalism. Nevertheless, the central premise of the 'varieties of capitalism' school is 'the persistence of differences in the organization of national political economies' and 'how resilient these institutional characteristics have been in the face of new political and economic challenges since the 1980s'.[22] The dominant thrust of scholarship – the three models approach – also focuses on difference. It insists that while France and other state capitalist countries in Europe have indeed been forced to make particularly important changes of late, the significance of the state in the economic arena remains distinctive enough to represent a clear difference in kind.

The very real economic differences between European nations persist despite facing similar pressures in the past quarter-century. Each state has sought to become more competitive, has become increasingly influenced by the power of financial markets that require quarterly profits, and has attempted to impose budgetary restraint. Each state has privatised industries, shied away from certain forms of national economic planning, and liberalised labour markets. But the pattern of changes does not support a thesis of economic determinism that all such changes lead inevitably to market capitalism. Long-standing directions in policy remain. Even when one factors in the force of common EU policy, there is little evidence to suggest convergence in economic policy, practice or even discourse. 'National practices remain differentiable into at least three varieties of capitalism despite . . . common economic challenges', argues Vivien Schmidt. 'European countries have not only followed different pathways to adjustment, they are likely to maintain these differences into the future.'[23]

The point here is that a paradigm that portrays Europe as Venus cannot account for the stark differences on such basic matters as economic orientation that exist between the many nations comprising Europe. If one wishes to counterpoise Europe and America, one must address the thorny issue of which Europe one has in mind. Is the European economy based on the managed capitalism of Germany, in which the big banks establish industrial policy while industrial relations represent a corporatist system of cooperation between organised labour and organised employers? Is Europe state capitalist like France, in which the state has retreated in recent years but still plays a dominant role in economic development? Or does Europe reflect Britain, which under Thatcher took dramatic steps down the road toward market capitalism? Comparisons based on simplistic dualisms become even more problematic once we move from the relatively abstract realm of economic models to specific sets of programmes such as those for social provision.

Social welfare: divergence or convergence?

Few contemporary questions are as contentious or as important as the future of the welfare state. And few issues, at least on the surface, lend as much credence to the popular paradigm. Western Europe has long been depicted as the birthplace of modern welfare, the effort to shield citizens from the vagaries of the market, while the US is routinely characterised as a welfare 'laggard', a 'deviant' case, where European-style welfare programmes arrived late and remained half-heartedly implemented.[24] It is well known that northern European and Scandinavian states spend in the neighbourhood of twice the amount (as a percentage of GDP) the US spends on social provision. And some writers are reluctant to label the US a welfare state at all, preferring to use terms such as 'semi-welfare state', 'workfare', 'residual welfare state', or 'work and relief state'. In no other policy area is the difference between Europe and the US so abundantly clear – a Social European world of cradle-to-grave protection for all citizens, and an American world that stigmatises and fails to provide for victims of the market economy.

For our purposes several interrelated questions are uppermost. To what extent is it useful to conceptualise social welfare policy differences with the popular paradigm? And, given the similarity of pressures facing all welfare states, is the US style of welfare becoming, or likely to become, the dominant style of welfare in Europe as well? In other words, to what extent has there been or will there be convergence in welfare provision between the US and Europe, and is globalisation (or other factors) forcing Europeans to abandon their generous approach to welfare and adopt more parsimonious US-style programmes? Certainly there is no dearth of literature suggesting that convergence toward the US model is inevitable unless Europeans consciously band together to bolster the idea of Social Europe. 'Global mobility of capital and production in a world of open economies have made the central policies of European social democracy unworkable', it has often been claimed.[25] Anne Gray is equally convinced that Europe is being relentlessly pulled toward American-style 'workfare', a process that will continue unless Europeans 'focus on the growth of collective services that will be truly collective and non-capitalist'.[26]

Defining and categorising welfare systems is a notoriously difficult task, but one fact is beyond dispute: the idea of a unified or even coherent European approach to welfare is a myth. There is no single European style of welfare that can be compared to the US style. The popular reasoning is at best misleading in regard to welfare, and at worst it is crude and disingenuous. Each nation in Europe has its unique history of welfare provision, and those differences remain abundantly evident today. Numerous and competing scholarly attempts to cluster the individual approaches to welfare in Europe serve only to highlight the social welfare mosaic that exists on this side of the Atlantic. Even if we look only at Western Europe and the Scandinavian states, one scholar insists that there are four clear 'social policy regimes', regimes 'so contradictory that an organic merging from below is not possible and "harmonisation" will necessarily have to come "from above"', namely the EU.[27] Other scholars are even more circumspect and warn that even by clustering European welfare regimes we run the risk of flattening important national differences. 'Even between the three Scandinavian countries, which are usually thought to be very similar' in their welfare policies, suggest Rothstein and Steinmo, 'there are significant differences in important policy areas.'[28] Yet models of welfare can be helpful. One of the most often-used models is that of Esping-Andersen, who posits 'three worlds of welfare capitalism' – liberal, conservative and social democratic – and who finds that all three 'regimes' exist simultaneously in Europe.[29]

Answering the second question – to what extent there is convergence toward the American system of welfare – can best begin with a brief look at the latest scholarship on several key states and regions. For the US itself, recent scholarship undermines the idea of 'exceptionalism' which has long been used to characterise its welfare policies. The popularity of typologies such as that of Esping-Andersen has served to situate US welfare on a spectrum of possibilities, rather than simply focusing on its departure from a mythic European norm. Esping-Andersen described the US as a liberal welfare regime, which is characterised by relatively low levels of social expenditure, means-tested benefits, and comparatively high levels of income inequality and relative poverty. Some authors continue to insist on the exceptional nature of the US system,[30] but most authors today challenge the idea of the US as 'laggard'.[31]

Certainly, the much-discussed and severe retrenchment of welfare programmes that began under Reagan and continued under Clinton, whose 1992 pledge to 'end welfare as we know it' epitomised the cutbacks, has reignited the debate about exceptionalism. In particular, the Personal Responsibility and Work Opportunity Act (1996) – by drastically reducing the number of those receiving direct welfare payments, driving many into low-paid and insecure work, and shifting much welfare responsibility from the federal to state governments – led some to suggest that the US can no longer be described as a liberal regime in Esping-Andersen's spectrum. And it is true that 'the retreat from social welfare has gone further and faster in the US than elsewhere'.[32] Yet whether recent policy shifts represent a lasting departure for the US, and whether European states are following this path, remains to be seen. Social expenditure as a percentage of GDP has increased in the US despite attempts to retrench. Some groups fared badly, to be sure, but 'the much-heralded cuts of the early 1980s were rather quickly reversed' and 'the widely-held view that the US has simply retrenched its spending on the poor does not entirely fit with these facts'.[33]

The question of whether Britain has been following the lead of the US is a hotly debated subject, with conflicts over the size, speed and nature of changes that have taken place since the Thatcher years. Britain's welfare system has always been difficult to characterise. In Esping-Andersen's schema, for example, it exhibited aspects of both the liberal welfare and social democratic regimes. In particular, Britain's mass provision of low-rent social housing and the universalist National Health Service did not fit the liberal regime model. Without question Britain has moved toward a more liberal regime approach on labour market and income maintenance aspects of social policy, and it has certainly made efficiency a byword in its rhetoric about welfare, leading some scholars to mourn the advent of a contemptuous 'Blair–Clinton orthodoxy' on welfare issues.[34] But the British welfare system continues to be a hybrid, representing a 'third way' between collectivism and unfettered free markets. Moreover, recent scholarship does not suggest a major lurch to the right in the foreseeable future. In power since 1997, the Labour government of Tony Blair has been 'keen to distance themselves from the anti-state, private market rhetoric of the new right'.[35] Rather than being forced to adopt a US-style system, the British have adopted, as one scholar put it, 'welfare with the lid on'.[36]

Germany's welfare state has often been lauded as one of the most stable in Europe. It is an example of what Esping-Andersen dubbed a conservative welfare regime, with wide and usually generous coverage, strong support for the traditional family structure, and benefits that are earned (usually through employment) rather than universal. Another important feature is that many programmes are funded by payroll taxes by employers rather than by general tax revenues and many are managed by a corporatist tripartite structure (employer, employee and the state) rather than directly by the state. This system has worked well in the German context, but like all welfare states, the stable German system is now in a state of crisis. In addition to facing pressures common to all welfare states, German stability is jeopardised by the enormous challenges created by unification in the early 1990s and the resulting huge transfers of resources to what was East Germany. Out of a total federal state budget of DM 412 billion in 1991, for example, 93 billion was spent on the direct costs of unification, and costs will continue to impact Germany well into the future. In Germany 'the welfare state has been made a culprit for many of the country's economic ills', and 'practically every programme has seen some change'. But there is no evidence that Germany has or will follow the American path. There has been no deviance from the unique German system. 'If anything there has been a tendency to adhere even more closely to existing principles when carrying out reform.'[37] Comparing the recent experience of German and US welfare regimes, one can see that 'their fundamentally different structures have led to distinct challenges and thus to completely different results'.[38]

The Scandinavian welfare states are by most measures the most generous in the world, and their universalist programmes are also admired for their egalitarian outcomes on questions of class and gender. The Scandinavian system, which represents the best examples of what Esping-Andersen calls the 'social democratic regime', is unlike the liberal or conservative regimes in that it is redistributive and designed to ameliorate conditions of poverty. In this the Scandinavians have been quite successful. In 1995, for example, Sweden's poverty rates (defined as the percentage of individuals with incomes below 50 per cent of the median) were just 6.4 per cent, while in Germany the rate was 9.1 per cent and in the US 17.1 per cent.[39] Beginning in the 1980s, these welfare regimes also began to retrench. This not only reflected leaner times economically, but also the idea that the lavish universalist programmes had 'grown to limits'.[40] Yet despite a greater focus on efficiency, some cutbacks, and a greater willingness to experiment with private initiatives in health care and social insurance, there has been no willy-nilly dismantling of programmes, no slash-and-burn to reproduce anything approximating an American approach. 'The basic structure of welfare systems has been preserved', observes Stein Kuhnle. The future might hold a mildly 'less generous welfare state' in Scandinavia, but the contours will remain the same.[41]

Looking at European welfare states collectively, the case for continued divergence in welfare provision is even clearer. Despite facing many similar pressures discussed above, and despite a general agreement that budgetary restraints are increasingly important and that welfare programmes must be rationalised, nation states continue to spend ever increasing amounts on welfare provision, and they spend it in different ways. 'While this logic of downward convergence is compelling, empirical analysis does not support it', writes Cynthia Kite. 'Neither levels of government revenue and expenditure, nor how governments spend are converging.'[42] When reforms are made, the type and extent of reform does not follow a predetermined 'race to the bottom' economic logic, but rather is in line with the nation's political culture and national attitudes about the role of the state and acceptable levels of taxation. Continuing national differences in social welfare provision show no signs of dissipation and there is no indication of convergence. Globalisation 'does not place systematic pressure on larger welfare states to 'run to the bottom' or 'run to the middle ground' of more market-conforming social policy', concludes Duane Swank.[43]

The key to understanding changes in welfare systems is to focus on both differences and similarities between unique national programmes, rather than to postulate some mythical and unified Europe. 'While there are broad, important and sometimes quite serious problems facing all modern welfare states', argue Bo Rothstein and Sven Steinmo, 'the varying characters and structures of the welfare states themselves critically shape the very definition of these problems as well as the likely solutions to them.'[44] John Myles and Paul Pierson offer two compelling reasons for the persistence of diversity among welfare states. First, whatever their politics, policy-makers 'are constrained by institutional and programmatic designs inherited from the past'. In other words, reforms are path dependent. More fundamentally, continued divergence is guaranteed by the same factor that gave rise to the original differences among welfare states – national politics.

> Cross-national differences in the organization and political capacities of the key constituencies affected by welfare states – workers, employers, women, private insurers, and public officials – continue to have an important impact on the character of reform, just as they have been in the past.[45]

Ahistoricism

Casting aside the bulk of recent scholarship on economics and the welfare state, adherents of the popular paradigm also play fast with history. Rather than seeing policies, programmes and political rhetoric in a constant state of change, they tend to portray the state of things today as having somehow reached a permanent condition, static and pre-determined. America was provoked into two world wars by European bellicosity. But now Europe, we are told, has learned its lesson and opted for a Kantian 'perpetual peace'. How things have changed. But will they never change again? Will America always be locked in militaristic mode and will Europe now and forever be the vanguard of negotiation and peace? History suggests not. The long view reveals nations in continuous flux, nations that do not conform to simplistic dualisms.

The inevitability of change is equally apparent in matters of political economy. Before World War II US social spending was greater than it was in Europe. In 1938 the federal government spent 6.3 per cent of GDP on social programmes such as employment insurance and public employment. At that time Sweden (3.2 per cent), France (3.4 per cent), Britain (5.5 per cent), and Germany (5.6 per cent) were the laggards. Nor has the Republican Party in the US always been as averse to welfare provision as it appears to be at present. Indeed, it was under Nixon that social spending in the US overtook military spending for the first time. Nixon even pressed for a national health insurance scheme and a guaranteed minimum annual income for all citizens. Policies and priorities change in Europe too, often quite rapidly. It was the French socialist Mitterrand who, in 1983, abruptly abandoned his expansionist neo-Keynesian monetary policies and generous social provision in 'the great U-turn', liberalising the money market and significantly curtailing social expenditure. Nor were the Scandinavian states always the frontrunners among nations in terms of social provision. As late as 1960 Norway's social spending as a percentage of GDP was roughly equivalent to that of the US (7.8 and 7.3 per cent respectively), and as late as 1983 its spending levels were far below those of Germany (19.0 to 24.4 per cent respectively). The only clear truth about political economy is that things change, and those paradigms that do not reflect change are untenable.

Conclusion

The paradigm of Europe as Venus is a poetic image. At its best it is an image of what might be: a continent in which the productive capacity of capitalism has at last been harnessed for the greater good, in which all citizens are guaranteed a decent standard of living regardless of market conditions. In the arena of political economy, the paradigm was originally designed to facilitate the realisation of this vision. Because of its erroneous judgements on globalisation, the autonomy of the nation state, and the question of European identity, and its disregard of history, it has now become an obstacle. At its heart, it is a romantic and not a realistic vision.

A truly Social Europe is possible but only if we are prepared to see beyond simplistic dualisms. An 'us and them' approach is counterproductive. The US and a richly diverse Europe must be seen as occupying different points on the same complex spectrum of policy choices, choices that result from national conditions and are conditioned by national politics. If common ground is to be found for Europe, if a Social Europe is to be created, it must not be based on a sense of anti-American superiority or the pretence that unity on important questions already exists. Common ground must begin with the recognition that there are no 'non-negotiable European realities'. With the entire European project now in disarray after the French and Dutch referendums of 2005, the advocates of Social Europe have a chance to base their dream on a more solid and realistic foundation.

Notes

1 The author would like to thank Dr Duncan Campbell and Dr Neil Campbell for their comments and suggestions on early drafts of this chapter.
2 Robert Kagan, *Paradise and Power: America and Europe in the New World Order* (London: Atlantic, 2003), p. 3.
3 Timothy Garton Ash, *Free World* (London: Penguin, 2005), p. 56.
4 Ibid., p. 271.
5 Jürgen Habermas and Jacques Derrida, 'February 15, or, What Binds Europeans Together: Plea for a Common Foreign Policy, Beginning in Core Europe', in Daniel Levy, Max Pensky and John Torpey (eds) *Old Europe, New Europe, Core Europe: Transatlantic Relations after the Iraq War* (London: Verso, 2005), p. 11.
6 Barbara Ehrenreich, *Nickle and Dimed: Undercover in Low-Wage America* (London: Granta, 2002), p. 214.
7 Allan Cochrane, John Clarke and Sharon Gewirtz, *Comparing Welfare States: Family Life and Social Policy* (2nd edn) (London: Sage, 2002), p. 147.
8 John Gray, *False Dawn: The Delusions of Global Capitalism* (London: Granta, 1999), p. 231.
9 Robert Went, *Globalisation: Neoliberal Challenge, Radical Responses* (London: Pluto Press, 2000), p. 28.
10 John Gray, *op. cit.*, p. 80.
11 Michael Lind, *The Next American Nation: The New Nationalism and the Fourth American Revolution* (New York: The Free Press, 2005), p. 203.
12 Phillip Cerny, 'Globalisation and the Changing Logic of Collective Action', *International Organisation*, 49, 1995, p. 621.
13 Jonathan Steele, 'A New Kind of Europe', *Guardian*, 15 June 2005; Will Hutton, 'My Problem with Europe', *Observer*, 5 June 2005.
14 Paul Pierson, 'Post-Industrial Pressures on the Mature Welfare States', in Paul Pierson (ed.) *The New Politics of the Welfare State* (Oxford and New York: Oxford University Press, 2001), p. 99.
15 Gosta Esping-Andersen, *The Three Worlds of Welfare Capitalism* (Cambridge: Polity Press, 1990); Evelyne Huber and John D. Stephens, 'Welfare State and Production Regimes in the Era of Retrenchment', in Pierson, *op. cit.*
16 Pierson, *op. cit.*, p. 99.
17 Torben Iversen and Anne Wren, 'Equality, Employment and Budgetary Restraint: The Trilemma of the Service Economy', *World Politics*, 50/4, 1998, pp. 507–46; Cynthia Kite, 'The Stability of the Globalized Welfare State', in Bo Sodersten (ed.) *Globalisation and the Welfare State* (London: Palgrave, 2004).
18 Geoffrey Garrett and Deborah Mitchell, 'Globalisation and the Welfare State', *European Journal of Political Research*, 39/2, 2001, pp. 145–77.
19 Paul Doremus, William Keller, Louis Pauly and Simon Reich, *The Myth of the Global Corporation* (Princeton: Princeton University Press, 1998).

20 Vivien Schmidt, *The Futures of European Capitalism* (Oxford and New York: Oxford University Press, 2002), p. 24; Fritz Scharpf, 'Economic Changes, Vulnerabilities, and Institutional Capabilities', in Fritz Scharpf and Vivien Schmidt (eds) *Welfare and Work in the Open Economy. Vol. I: From Vulnerability to Competitiveness* (Oxford and New York: Oxford University Press, 2000).
21 Gray, *op cit.*, p. 92.
22 Stewart Wood, in Peter Hall and David Soskice (eds) *Varieties of Capitalism: The Institutional Foundation of Comparative Advantage* (Oxford and New York: Oxford University Press, 2001), p. 247.
23 Schmidt *op. cit.*, p. 303.
24 Seymour Martin Lipset, *American Exceptionalism: A Double-Edged Sword* (New York: Norton, 1996).
25 John Gray *op. cit.*, p. 89.
26 Anne Gray, *Unsocial Europe: Social Protection or Flexploitation?* (London: Pluto, 2004), p. 193.
27 Stephan Leibfried, 'Towards a European Welfare State?', in Christopher Pierson and Francis Castles (eds) *The Welfare State Reader* (Oxford: Polity, 2000), p. 190.
28 Bo Rothstein and Sven Steinmo, 'Restructuring Politics: Institutional Analysis and the Challenge of Modern Welfare States' in Bo Rothstein and Sven Steinmo (eds) *Restructuring the Welfare State: Political Institutions and Policy Change* (New York: Palgrave, 2002), p. 3.
29 Gosta Esping-Andersen, *The Three Worlds of Welfare Capitalism* (Cambridge: Polity Press, 1990).
30 Frank Dobbin, 'Is America Becoming More Exceptional? How Public Policy Corporatized Social Citizenship', in Rothstein and Steinmo, *op. cit.*
31 Edwin Amenta, *Bold Relief: Institutional Politics and the Origins of Modern American Social Policy* (Princeton: Princeton University Press, 1998); Theda Skocpol, *Protecting Soldiers and Mothers: The Political Origins of Social Security in the United States* (Cambridge, Mass.: Harvard University Press, 1992).
32 Allan Cochrane, John Clarke and Sharon Gewirtz, *Comparing Welfare States: Family Life and Social Policy* (2nd edn) (London: Sage, 2002), p. 148.
33 John Clarke and Frances Fox Piven, 'United States: An American Welfare State?' in Pete Alcock and Gary Craig (eds) *International Social Policy* (New York: Palgrave, 2001), p. 40.
34 Bill Jordan, *The New Politics of Welfare* (London: Sage, 1998), p. 2.
35 Pete Alcock and Gary Craig, 'The United Kingdom: Rolling Back the Welfare State?' in Alcock and Craig, *op. cit.*, p. 139.
36 Howard Glennester, 'Welfare with the Lid On', in Howard Glennester and J. Hills (eds) *The State of Welfare* (2nd edn) (Oxford: Oxford University Press, 1998), p. 55.
37 Mary Daly, 'Globalisation and the Bismarckian Welfare States', in Robert Sykes, Bruno Palier and Pauline Prior (eds) *Globalisation and European Welfare States: Challenges and Change* (New York: Palgrave, 2001), p. 96.
38 Elgar Rieger and Stephan Leibfried, *Limits to Globalisation* (Oxford: Polity, 2003), p. 188.
39 Cochrane, Clarke and Gewirtz, *op. cit.*, p. 202.
40 John Stephens, 'The Scandinavian Welfare States: Achievements, Crisis and Prospects', in Gosta Esping-Andersen (ed.) *Welfare States in Transition: National Adaptations in Global Economies* (London: Sage, 1996).
41 Stein Kuhnle, 'The Nordic Welfare State in a European Context: Dealing with New Economic and Ideological Challenges in the 1990s', in Stephan Leibfried (ed.) *Welfare State Futures* (Cambridge: Cambridge University Press, 2001), p. 394.

42 Cynthia Kite, 'The Stability of the Globalized Welfare State', in Bo Sodersten (ed.) *Globalisation and the Welfare State* (London: Palgrave, 2004), p. 213.
43 Duane Swank, *Global Capital, Political Institutions, and Policy Change in Developed Welfare States* (Cambridge and New York: Cambridge University Press, 2002), p. 275.
44 Rothstein and Steinmo, *op. cit.*, p. 3.
45 John Myles and Paul Pierson, 'The Comparative Political Economy of Pension Reform', in Pierson, *op. cit.*, p. 306.

13

NEW WORLD AND OLD EUROPE

American screen images of Britain and France

Beatrice Heuser

In 2003, on the eve of the joint US and British invasion of Iraq, US Secretary of Defence Donald Rumsfeld famously criticised France, Germany and Belgium for opposing this operation by calling them 'old Europe', by contrast praising as the 'new Europe' those countries which supported that US policy – above all Britain, but also Spain, Poland and some other East European countries. Many Europeans did a double-take at this characterisation, which obviously had a number of implications. To some extent, 'new Europe' clearly meant countries that had recently been won over to liberal democracy and free market economics, a categorisation which fits Spain and the countries east of the former Iron Curtain. But 'new Europe', in Rumsfeld's definition, also included Britain, a country whose successive governments had stood shoulder to shoulder with the US on most major security policy issues since the Second World War. The implication was also that Britain was a country whose present government clearly understood the new rules that in US opinion dominated the 'game' now that the Cold War was over, and that it was among European countries inclined furthest towards the US example, backing economic competitiveness by reducing social security, refusing to adhere to the EU's social charter and the Euro currency. Rumsfeld's statement was a deliberate snub to those in France and Germany who think the EU the more advanced form of social and economic organisation, and particularly to French modernism.

Yet American perceptions of Europe are more complex than the mere division into old and new, 'progressive' (and therefore supportive of US foreign policy) and 'reactionary' (that is, critical of George W. Bush). To explore further the relationship between America and Britain on the one hand, and America and France in particular on the other, it may be necessary to leave the realm of strictly rational, measurable interpretations, and venture into the realm of subjective perceptions, gut feelings, myths and folklore. I want to open or attune your minds to the larger world of fantasy and image, of interpretational cliché and literary genre in order to follow my arguments. I shall now appeal to your inner eye, and abduct you into the world of cinema.

Just before we set out, I shall throw in the rationale that others have used before me: suggesting that images and myths add a dimension to our understanding of international relations which cannot be captured by any of the purely 'rational' or organisational explanations. One of them is Keith Shimko, who argued:

> Images can be conceptualised as mental pictures composed of our accumulated knowledge, that is, beliefs and attitudes regarding our surroundings. They are intellectual constructs that bring order to the world. . . . Since images are mental constructs developed through socialization and experience, they can be altered in response to new information.

Moreover, 'Our images of certain objects may be different tomorrow, and the changes in these images may produce important behavioural modifications. Thus, characterising images as significant and fairly stable mental artefacts that help determine our actions does not imply that they are constant and immutable.'[1] This sense of mutability is reinforced by Philip Burgess, who wrote: 'an image is a dynamic concept: that is, images are subject to redefinition as a result of the operation of interactive and feedback processes'.[2] John Steinbruner sees 'an oscillation over time between a number of belief patterns',[3] an expression that I, too, prefer over world view, as the latter implies a sense of coherence that rarely exists.[4] In the following, therefore, I shall try to shoot at the moving picture (in more senses than one) which Britain represents to American spectators, contrasting it with that of France, and we shall see that old is not necessarily old, and new not always new.

The Black Knight

The Black Knight (2001) stars Martin Lawrence (a black American born in Germany) who was also the executive producer of the film. It belongs to the genre of novels and films that deal with the American experience of the Old World, more precisely England, here quite literally transposed into the Middle Ages. Lawrence plays the hero, Jamal Walker, who works in a menial job in the family leisure centre *Medieval World*, somewhere in the Boulevard area of Los Angeles, on the corner of Florence and Normandy Boulevards. It is a little run down. The staff are all black, and the impressive manager, a weighty lady, prides herself on having created and provided 'quality jobs' for the (black) community for twenty-seven years. A threat has arisen, however: nearby, a rival theme park has been set up, called *Castle World*. The manager of *Medieval World* announces this to her staff with the comment that they must all now try harder than before to make their business pay off. Jamal, who comes across as a light-hearted but also irresponsible charmer, suggests to their boss that she should sell out and retire comfortably to a beach – to which she reacts with disdain, sending Jamal off to clean the moat. Jamal accidentally falls in and loses consciousness. When he comes to, he is in a different world, a different century, which he takes to be the land of the (indeed very convincing) theme park *Castle World*.

It turns out that with all his brashness and light-heartedness, Jamal is a good guy. He starts by saving the life of a run-down knight who has turned into an alcoholic as he has failed to serve his queen well. Jamal takes pity on the knight, gives him a couple of dollar bills, and advises him, first, to buy himself some Ikea furniture for his makeshift tent, but then, to get food stamps and support from the social services, or indeed to move into a shelter. Then Jamal ambles on and finds himself in front of a very impressive castle. He is nearly ridden down by a number of knights, and confronts one of them, having penetrated into the confines of the walls as his identity is mistaken to be that of a long-awaited messenger from Normandy. He pours out line after line of political correctness when taking on the leading knight, who unsurprisingly turns out to be the arrogant villain, and who sexually harasses the ladies-in-waiting in the castle. 'Do you mock me, Moor?' is met with 'Hey why don't you take a chill-out pill?', and Jamal is appalled to find that the 'employees' of what he thinks is 'Castle World' call their manager 'king', something he thinks they should take up with their union.

A Connecticut Yankee

The genre to which *The Black Knight* belongs was invented by Mark Twain, who in 1889 published *A Connecticut Yankee in King Arthur's Court*. This burlesque contrasted the knightly ideals and behaviour at King Arthur's court (as known from medieval romance) with the 'enlightened', democratic but ultimately inhumane capitalism of late nineteenth-century America. The Yankee of the title and narrator is one Hank Morgan, probably named after the great financier J.P. Morgan, who like Twain's hero was from Hartford, Connecticut. Hank, a worker in a factory there, is knocked out during a scuffle resulting from a strike, and regains consciousness in Camelot. He is recognised and feared as an outsider, and is sentenced to death by burning. He is saved by claiming magical powers, which he demonstrates in predicting a solar eclipse. Thought to be a magician, he thenceforth impresses the British by using dynamite to blow up a tower belonging to King Arthur's and his antagonist, Merlin. He then sets out to modernise Arthur's Britain, supplying his knights with bicycles, firearms and other modern weaponry, which on one occasion leads to the massacre of 25,000 knights in an American

Civil War-type battle. Medieval ritual and sensitivities are brushed aside with brash 'rationalism' and common sense. But in the end, Merlin wins, by casting a spell on Hank that puts him to sleep, from which he awakens again back in his own time, in his own country and town. Twain's wit lashes out in all directions. The novel is not only a satire on Victorian class-obsessed society, but also a fierce criticism of ruthless American capitalism, as Hank Morgan uses every opportunity to make money out of the innovations he introduces into the Old World.

Twain's book became a highly popular inspiration for the American film industry. The first film version appeared in 1920, directed by Emmett J. Flynn. It was followed by productions by David Butler in 1931, Tay Garnett in 1948 (starring Bing Crosby), and Mel Damski in 1989, each appearing with the actual title of Twain's novel (although in Europe, they were often shown with the omission of 'Connecticut', as *A Yankee in King Arthur's Court*). A Soviet director also took up the theme in 1989, under the title *Novye Prikluchenia Janke Pri Dvore Kovola Artura* (The New Adventures of a Yankee at King Arthur's Court). A host of other novels and films were inspired by the story. In 1979, Russ Mayberry directed *The Spaceman and King Arthur* for Disney. Other films took up the 'Yankee in . . .' theme. The first, produced in 1938, was based on A.P. Garland's eponymous book *A Yank at Oxford*, published in the same year, and was remade in 1984 as *Oxford Blues* starring Rob Lowe. The original spawned a series of films: *A Yank in the R.A.F.* (1941), *A Yank at Eton* (1942), *A Yank in London* (1946), all in a friendly way commenting topically on the culture shock experienced by Americans whom the Second World War had brought to Britain. The mocking, condescending Yank did not merely visit medieval and contemporary Britain, but also Ancient Rome, and other parts of the globe such as Libya, Burma (both during the Second World War) and Indochina (in 1952, well before America's Vietnam War). Laurel and Hardy, perhaps a little surprisingly in 1940, in their endeavours to come to Oxford encounter rather arrogant British fellow-students playing pranks in *A Chump in Oxford.* The theme also resonates in the two films *An American Werewolf in London* (1981), written and directed by John Landis, and its sequel, *An American Werewolf in Paris* (1997), directed by Anthony Waller, in which American tourists are attacked and bitten by werewolves in the respective capitals of Britain and France, and this is taken as an excuse to point out the beauty and romanticism of landmarks and tourist attractions, padded in copious self-mockery and enhanced with great special effects.

Twain's leitmotif of the modern man (or American) who baffles 'savages' (or old-fashioned Britons) with his scientific knowledge was appropriated by Hergé, the Belgian author of *Tintin et le temple du Soleil* (1946) where the young hero also escapes his own execution by South American Indians by predicting a solar eclipse. We also find self-critical British appropriations of the theme in a film called – coincidentally – *The Black Knight* (1954), featuring the actor who the previous year had starred in *Shane*. In this version of Twain's story, a modest swordsmith unmasks a traitor at the court of King Arthur. As one critic wrote, 'Alan Ladd galahads with wild west *gentilesse* in this Technicolored rampage through British history.' In all of these films, Hollywood both makes fun of a backward England and spreads information about it – at least in the form of national stereotypes – across the globe. But in fact Hollywood does more than mock the English – it also positively propagates knowledge about at least one English hero.

Robin Hood

If *The Black Knight* belongs to the genre inspired by Mark Twain, its principal character, Jamal, is also representative of a hero of British folklore, Robin Hood. Having encountered the outlaws, who support the True Queen against the Usurper King, but are in want of a leader, Jamal organises them and ensures the crucial charge on the Usurper King's castle that leads to his overthrow and to the reinstatement of the True Queen. Before his arrival, the outlaws have lost heart: their chief military leader, the knight we have already encountered, has turned into a drunkard, and their Queen has failed to rouse them, having attempted to do so with a (Thatcheresque?) speech in which she merely drawls the word 'Eeengland, Eeeeeeeengland' over and over again. In assuming the leadership of this rabble, Jamal steps into the mythical shoes of Robin of Sherwood, who defended the rights of the true king Richard the Lionheart against his brother, John Lackland. Jamal wins over the hearts and minds of the despondent rebels, rekindling their courage by appealing to ideals greater than mere loyalty to one prince: the interest of the society as a whole.

This legend strikes a particular chord in the American psyche. To quote from a speech given by Jeane Kirkpatrick, when US ambassador at the UN, to an American audience:

> We are the inheritors and the embodiment, the twentieth century American version, of a long struggle against arbitrary power. This struggle has characterised our tradition, the Western tradition, since long before that reluctant King John was persuaded to sign the Magna Carta in the early thirteenth century. We are the heirs and bearers of a tradition that was shaped by the persistent efforts of Englishmen to limit the power and jurisdiction of their kings.[5]

The story of Robin Hood appeals to Americans in several ways. Robin Hood stands up for the people in opposition against dictatorship and tyranny. He is also the archetypal social reformer, taking from the rich, giving to the poor. Robin Hood stands for one important strand of British social and political history, which in the common narrative is traced from his own legend starting around 1200, on through the struggle of the barons against King John a few years later, to the Peasants' Revolt under Wat Tyler in the late fourteenth century, to the Tudor Rebellions, and the English Civil War (which led to the establishment of a form of republic in the Commonwealth of Oliver Cromwell), then through the Corn Law demonstrations and the machine breakers in the nineteenth century, and on to socialism, the Labour Party and the National Welfare State. Robin Hood is thus seen – and not only in Britain – as the protagonist of social justice and good governance and an early fighter for freedom and democracy. In American schoolbooks of the nineteenth and early twentieth centuries, the 'bold outlaw, Robin Hood' is praised as the subject of medieval ballads and romances, and is part of the pantheon of heroes whose contributions to humanity are celebrated. In *How Civilization Grew in the Old World and Came to the New*, an American schoolbook written by Smith Burnham at the end of the First World War, the legend is traced under the heading, *Our Beginnings in Europe and America*.[6]

Four years after the end of that war, in which America had come to the defence of its allies in Western Europe much as Robin Hood, the nobleman of Norman stock came to the aid of the oppressed Anglo-Saxon native population in England, *Robin Hood* first strode silently across American movie screens. He was equipped with a voice and Technicolor when he returned swashbucklingly in 1938, as dictatorship once again oppressed Europe. In this film, one of the very few four-star movies listed in *Halliwell's Film Guide*, the villains are, oddly, the Normans, and the heroes (quite ahistorically, including Robin Hood, who himself was of Norman stock) are the Anglo-Saxons. Walt Disney took up the theme in 1952 with a feature film, produced with British money. In 1973, Disney also produced the animated version, a film featuring a fox as Robin Hood and cheerful little animals in all other parts, which has, as usual with Disney, given birth to a big industry of comic strips, T-shirts and mugs with motifs from the film.

The first spoof on the subject was Danny Kaye's brilliant *Court Jester* (1956), in which a witty but not very courageous man accidentally becomes the key player in the restoration to the throne of the rightful king against his wicked uncle who wanted to have him killed. Danny Kaye is in fact the jester in the camp of a Robin Hood character (the Black Fox). The real king is a baby whose claim to the throne is proven by a birth-mark on his bottom, which is ritually exposed to all his subjects, who curtsey and bow appropriately upon beholding the royal posterior. Inspired clearly by the lightness of touch of this film and its 'Broadway Musical' character was a very charming and humorous TV series *Robin Hood*, which hit the screens in the 1970s.

By contrast, *Robin Hood: Prince of Thieves* (1991), an attempt to present the matter more seriously by casting Hollywood's most serious romantic actor, Kevin Costner in the title role, received mixed reviews, and gave rise to the second parody of the genre in Mel Brooks's *Robin Hood: Men in Tights* (1993). This film follows the Costner film scene by scene. A cheerful spoof of the whole genre also appears in the animated Dreamworks film *Shrek* (US, 2002), in which Robin Hood oddly has a French accent, and he and his men (in tights) break into dance to the Village People's 1970s hit 'YMCA'. Here Robin Hood does not come away as hero. Rather he and his men experience serious damage at the hands and feet of Princess Fiona, whose Kung Fu abilities they had not expected when lavishing unwelcome attentions upon her (perhaps Robin's French accent is supposed to evoke the cliché of the imposing Gallic Lover). All in all, Robin Hood is clearly an American household name (or should one say – *pace* Disney – a Nursery Name?) and hero, as demonstrated in a BBC World Service broadcast, *The World Today*, on 15 October 2003. The controversy was whether Robin Hood was actually from Nottinghamshire or instead from Yorkshire. The debate by historians arguing the two cases was ended with the intervention of an American: 'But c'mon', he interjected, 'we all know that Robin Hood is American!'

American images of Britain

The appeal of Robin Hood in America is, then, a clear manifestation of the cultural 'special relationship' between Britain and the United States. American schoolbooks, particularly in the late nineteenth and early twentieth centuries, made it quite evident that the US was a child of British culture, and they underscored the importance of the cultural link between the 'New England' that was founded by the Pilgrim Fathers and the 'Old England', which they had left behind, but whence they had brought their cultural heritage and their traditions, including both Puritanism and egalitarianism.[7] As Paul Johnson put it, America was seen as 'the posthumous child of the English revolution'.

Nevertheless, there is also a very strong self-definition of the US as having been created in opposition to Britain, which was after all for centuries a country of religious intolerance and colonial oppression, which ended its revolution with the restoration of monarchy, which carried its class system well into the twentieth century and which stood for all that America rebelled against. This in itself is strongly reflected in films on the subject of the American disengagement from British domination in the eighteenth century. Not surprisingly, the villains are the British in all film versions of Fenimore Cooper's *The Last of the Mohicans*, and their accents stand out increasingly strongly towards the end of the twentieth century. When *The Last of the Mohicans* was first filmed in 1936, Hollywood was full of actors who spoke a soft, only very slightly Americanised version of the Queen's English, side by side with scores of European *émigrés* with all sorts of weird and wonderful accents, which rarely stopped them from becoming great stars.

Differences in accent were thus not as pronounced yet, nor in the late 1940s, when the *Last of the Redmen* was filmed. Hosts of British actors migrated to America, to play parts which were frequently not nationally stereotyped.

Things changed very considerably in the last three decades of the twentieth century. All of a sudden regional accents penetrated into Hollywood, and ousted the English lilt of the East Coast Establishment. From then on, heroes in American films usually spoke with very distinct American accents, and villains increasingly with foreign accents. And it became seemingly most politically correct (or at any rate, tolerable) to cast English actors in the latter roles.

One would of course expect this in the 1992 film version of *The Last of the Mohicans* in which Daniel Day-Lewis plays Leatherstocking. Here the British with their English accents come across as particularly condescending, arrogant, inflexible and mean, just as they had in *Revolution* (1985), featuring Al Pacino, Donald Sutherland and, as the English villain, Steven Berkoff. However, it is less logical that the British-born American characters in these films speak with late twentieth-century American accents. One would expect to find the villains speaking with English accents in films like *Rob Roy* or *Braveheart*, both great celluloid tributes to America's citizens of Celtic descent, with the inevitable Anglophobia they have inherited. But logic does not necessarily dictate the English accents of the evildoers in films like *Waterworld*.

I have deliberately said *English* accents, because Scottish and Irish accents assume a quite different role in American films. The romantic, idealistic archaeologists in *Timeline* are played by two Scotsmen who help the considerate French against an attack from the wicked English in the Hundred Years' War, and Sean Connery's American film career was helped if anything by the hint of Edinburgh thunder in his voice. Shrek, significantly, speaks with a Scottish accent, Fiona with an American accent, and her arrogant parents as well as the selfish duke Humphrey with English accents. Everybody's darling in the film, the donkey, is voiced by Eddie Murphy with his notorious black Bronx accent – a subject to which we shall return presently.

The pattern of 'American – good, English – bad' has another dimension, the Celtic one. Even beyond the historic settings of *Braveheart* and *Rob Roy*, where this makes sense, American films usually cast Scots and Irishmen in the roles of heroes, reflecting pro-Celtic sentiments among many Americans who can trace their lineage back to nineteenth- and early twentieth-century migration from the British Isles to the US. Whether it was for religious reasons or because of Highland Clearances or the Potato Famine, the Celtic migrants from Ireland and Britain to the US brought with them a profound hatred of the English, which has become a strong feature in the relationship between both sides, reinforcing the image of the English as oppressive, exploiting and selfish colonialists.[8] Again, perceptions of this go back well into the eighteenth century. On 13 January 1772, Benjamin Franklin wrote from London to a friend,

> I have recently made a tour through Ireland and Scotland. In those countries, a small part of the society are landlords, great noblemen, and gentlemen, extremely opulent; living in the highest affluence and magnificence. The bulk of the people are tenants, extremely poor, living in the most sordid wretchedness, in dirty hovels of mud and straw, and clothed only in rags. I thought often of the happiness of New England, where every man is a freeholder, has a vote in public affairs, lives in a tidy warm house, has plenty of good food and fuel, with whole clothes from head to foot, the manufacture, perhaps, of his own family. Long may they continue in this situation!
>
> . . . I should never advise a nation of savages to admit of civilisation; for I assure you, that, in the possession and enjoyment of the various comforts of life, compared to [the Irish and the Scots], every [American] Indian is a gentleman, and the effect of this kind of civil society seems to be, the depressing of the multitudes below the savage state, that a few may be raised above it.[9]

Accordingly, Franklin called the English a 'haughty, insolent nation' (letter of 25 February 1775).[10]

Curiously, this strand of Anglophobia and of Celtophilia, so strongly reflected in the products of Hollywood, has hardly been influenced by the decades of loyalty (or, as some say, subservience) shown by British governments to the US in matters of foreign and defence policy. Even the First and Second World Wars, which otherwise are always cited as the great catalysts of Anglo-American bonding, could actually prove very alienating. On the one hand, the British, with their highly developed class-consciousness, could not understand the way soldiers treated their commanders and commanders their subordinates in the US Army – or so it seemed to the British who were less attuned to the more subtle status differentiations that operate in US society. The Americans, in turn, were put off by the universally acknowledged hierarchies in the British forces and society in general. An American pilot who joined the RAF at the very beginning of the Second World War out of sympathy for the cause of liberty noted shortly afterwards with disgust:

> I sometimes feel that England does not deserve to win this war. Never have I seen such class distinctions drawn and maintained in the face of a desperate effort to preserve democracy. With powers of regulation and control invariably centred in the hands of the few, and abuse and preservation of the old school tie . . . stronger than ever on every side . . . it has been well and truly said that

> General Rommel . . . would never have risen above the rank of NCO in the British Army. The nation seems inexplicably proud of the defects in its national character.[11]

Old England, New France

England, then, is seen in the US (and France) as a bastion of reactionary thinking, living and doing, which since the eighteenth century has opposed all revolutions, independence and the development of republics. 'New France and Old England' is one apt chapter heading in Norman Hampson's study of French perceptions of Britain during the French Revolution,[12] but it is just as much a pattern in American perceptions of England on the one hand and France on the other.

The French colonial empire in North America had been eliminated by the British by 1763, with the result that the French were not associated with European imperialism in the American master narrative of struggle for liberation against colonial oppression. Indeed, there were vague memories of France being on the opposite side from the British in the struggles of the eighteenth century, the constellation which led Louis XVI to take sides again against the British with the Americans fighting for independence. While the underlying rationale was that my enemy's enemy is my friend, this nurtured the belief in the vital quality of French support for the rebels, and the commonality of their cause. In the light of the influence of the *lumières* on American thinking about democracy, royal French balance-of-power politics retrospectively became transfigured mythically into a joint struggle for freedom, against royal tyranny. By the mid-nineteenth century, George Bancroft, a former US minister plenipotentiary at the courts of both Paris and Berlin, saw it as part of his mission to underline the vitally supportive part which the France of Louis XVI had played in the American battle for independence from the British motherland.[13]

This part of the grand narrative of American independence falls under the heading of the Lafayette myth. The Marquis de Lafayette supported the Americans in their war against Britain, and went on to play a part in the French Revolution. While the French see him as a vehicle for the transport of the ideas of the French *philosophes* to America, Americans like to see him as an exporter of American values to France. Either way, he is symbolic of the transatlantic bond in this era of Revolution, and it was after all France that was the first country to support the new United States with the treaty of 1778. Lafayette returned to America in 1824–5 for a tour of all twenty-four of its states, and was received as a popular hero, even then contributing to the early flourishing of the souvenir industry that produced plates and figurines as collectors' items.[14]

The impact of the Lafayette myth and the predilection for Paris of Thomas Jefferson resulted in the French capital becoming transfigured into not merely the fairy godmother of the newly born US, but the actual *mother* of American independence. Paris became not only the City of the *Lumières* but the 'City of Light'. Thomas Jefferson, while shocked by the 'physical and moral oppression' of 'the great mass of the people' in France (and elsewhere in Europe), wrote admiringly from Paris on 30 September 1785:

> Here, it seems that a man might pass a life without encountering a single rudeness. In the pleasures of the table, they are far before us, because, with good taste they united temperance. They do not terminate the most sociable meals by transforming themselves into brutes. I have never yet seen a man drunk in France, even among the lowest of the people. Were I to proceed to tell you how much I enjoy their architecture, sculpture, painting, music, I should want words. It is in these arts that they shine. The last of them, particularly, is an enjoyment, the deprivation of which with us, cannot be calculated. I am almost ready to say, it is the only thing which from my heart I envy them . . .[15]

A great many of the common American stereotypes about France which have survived happily until this day can already be found in this letter.

American images of France

Much is made in France of the commonalities of values of the American War of Independence and the French Revolution, both born of ideas of the Enlightenment. The treaty between France and America of 1778, the first concluded with a foreign power, and the first to recognise the United States as a sovereign entity, was commemorated in France in 1878 with the creation by the French sculptor Frederic Bartholdi of the Statue of Liberty, a French gift to New York. After the First World War had been jointly fought and won, the French philosopher Emile Boutroux wrote on 1 November 1918 in the *Revue des Deux Mondes:* 'If there have ever been two peoples in this world, who instinctively feel mutual attraction, they must be the French and the American peoples.' In January 1919 the writer Jean Guehenno wrote in the *Revue de Paris:* 'There is a spiritual bond uniting France and America. These two countries are at the forefront of human adventure. They are both driven by the same divine enthusiasm for liberty.'[16]

While the kinship of the American republic and the first French republic was present in the minds of leading Americans in the late eighteenth century, it plays a small role only in American cultural representations of France. The only literary-cum-cinematic stereotype taking the theme of oppression and exploitation followed by revolution and republic that has made it to Hollywood is that carried in the various versions of *Les Miserables*, based on the novel by Victor Hugo. In 1935, Twentieth Century Fox produced

Richard Boleslawki's Oscar-winning version with Frederic March, Charles Laughton and others. In 1952, Lewis Milestone directed the remake, which starred amongst others Michael Rennie and Debra Paget. A French version was produced in cooperation with Warner in 1975 with Jean-Paul Belmondo in one of his best roles. Bille August directed the American 1998 version (starring Liam Neeson), and this is the least interesting of the four. Bille August tried to cash in on the long-standing success of the Alain Boublil, Claude-Michel Schonberg and Herbert Kretzmer musical that had haunted London's West End and New York's Broadway for several years by then. Nevertheless, the idea of oppression overcome by revolution parallels that of the American struggle for independence from Britain, and thus appeals to the master narrative of the US, which probably accounts for the popularity of this European story on the other side of the Atlantic.

Other than that, stereotypes about France fall into three categories. The first set is clearly inherited from the British and reinforced, following Jefferson's experience, by the American experience of French *savoir vivre* and 'gay Paree' in the First and Second World Wars and the years following until de Gaulle threw NATO's headquarters and all foreign forces out of France (1967). Paris was the Mecca of American intellectuals caught up in the ferment of European politics in the 1920s, 1930s and 1940s, attracting such luminaries as Ernest Hemingway, Gertrude Stein, Henry James, Natalie Clifford Barney, and among many, the black writers Richard Wright and James Baldwin.[17]

The second set of stereotypes revolves in part around the cliché of good French gastronomy, complete with French waiters and French chefs, which is still an integral part of the menu of American comedies at the turn of the millennium. For example, both Disney's *Little Mermaid* (1989) and *Aladdin* (1992), and Dreamworks' *Shrek 2* (2004) feature a French chef or waiter. *Ooh-La-La Paree* as pinnacle of *savoir vivre* is indelibly linked with Gene Kelly's *An American in Paris* (MGM 1951), directed by Vincente Minelli. One Frenchman has summarised this set of stereotypes succinctly as consisting of 'Sexy Pigalles, mad artists, naked models, and businessmen yearning for US Dollars, expensive restaurants, delicious food and dangerous traffic.'[18] Indeed one must also include in this collection of images 'l'amour, toujours l'amour': the cliché of the French as the nation who are masters in the art of the ultimate enjoyment of love (but also of betrayal and adultery). The epitome of this cliché is the 1958 American MGM version of *Gigi*, starring Leslie Caron in the eponymous role. In the late 1950s and in the 1960s, Europe stood for more liberal mores, the sexual revolution, and the French side of this was incarnated by Brigitte Bardot, a double-initial answer to Hollywood's MM. Maurice Chevalier, who played rather popular working-class figures in French films, came to incarnate the French aristocrat, the ultimate playboy and lover, in American films: not surprisingly, he plays an aristocrat in *Gigi*, with his Lolita-ish song 'Thank Heavens for

Little Girls'. This stereotype persists: it is served up, indeed, in *The Black Knight*, when we hear of the new import (by the hero, Jamal) of the French kiss.

The last set of literary-cum-cinematographic stereotypes is that embodied in Alexander Dumas' *Three Musketeers*, like the legend of Robin Hood a popular theme in Hollywood. There is a parallel: like Robin Hood, the Three (in fact four) Musketeers show unswerving loyalty to the (good, real) king and try to combat the machinations and political intrigues of the regent (Cardinal Richelieu in the Three Musketeers; in the Robin Hood yarn John Lackland and the Sheriff of Nottingham). An Anglophobe strain no doubt appeals to Gallic and Gaelic-American audiences alike, when the English 'Milady' conspires to harm the French monarchy. The first American film of the Three Musketeers story appeared in 1935, then again four years later. Perhaps most successful was the 1948 version starring Gene Kelly as D'Artagnan, Lana Turner as Milady de Winter, and Vincent Price as Richelieu. In all these films, American cinema was the vehicle transporting French culture to movie screens throughout the world. Another popular American adaptation came in 1973 (also known as *The Queen's Diamonds*), directed by Richard Lester, and starring Michael York, Oliver Reed and Richard Chamberlain, as well as Faye Dunaway as Milady and Charlton Heston as Richelieu. This has a sequel, *The Four Musketeers* or *The Revenge of Milady* (1974, with the same director and cast). The cliché of the *bon vivant* Frenchman, eating, drinking and whoring, but actually hiding a deep *chagrin d'amour*, is brought across very powerfully in this version. Finally, Walt Disney came out with a new rendition in 1993 starring Mickey, Donald and Goofy, but this turned out a flop.

The Musketeer films obviously hark back to a romantic past. In the machinations of Richelieu and Milady, the archetypical cliché of a corrupt old Europe is revealed, but the Musketeers honourably combat it, just as the heroes in *Les Miserables* ultimately triumph over oppression. Therefore, ultimately, neither these nor any of the other film stereotypes of France show France as backward to the extent that England is depicted as decadent in American films.

Politically, however, France's popularity in America declined sharply once de Gaulle returned to power and founded the French Fifth Republic. The 'perfidious Albion' once embodied by 'Milady' increasingly gave way to France which under de Gaulle played a challenging role in NATO, regularly provoking American anger. From the 1960s until well into the twenty-first century, France and America could easily be called the worst friends in the world, as Jean Guisnel put it.[19] Interestingly, I can think of few if any effects of these political divergences on American cinema – paralleling, one might say, the relative absence of bows in the direction of British political loyalty to the US.

British and French perspectives on race

Let us return to our Black Knight. There is a deep irony in this film, in that it is a black American who sees (admittedly, medieval) Europe as crude and backward. For Europe was far ahead of the US in the emancipation of coloured people. Britain adopted legislation against slavery in 1807 for the UK, and in 1833 for the Empire. France banned slavery in 1794 for Saint Domingue, and in 1848 for the rest of the French colonies, thus preceding California, the first US state to declare itself without slavery in 1850. There is no mention of human equality in the American Declaration of Independence (Jefferson confined himself to declaring that 'all men are created equal'). The fact of racism is a strong theme in de Tocqueville's otherwise so admiring description of the United States. The condition of black slaves and native Americans filled him with pity and revulsion, for the difference between these human groups, as he argued, was clearly man-made and socially constructed. While the American Civil War was fought primarily over the issue of slavery, it hardly needs to be mentioned here that even the victory of the anti-slavery North did not resolve the issue. The rights of black citizens of the US were still heavily fought over as recently as the 1960s and 1970s.

This is not to say that British and French merchants were never involved in the slave trade – far from it, and many benefited hugely. Nor was racism absent from both nations' colonial policies, and as in other European societies, Social Darwinism took its ugly toll in both countries. And yet British and French racism was by and large something which happened overseas: until the years of decolonisation after the Second World War, there were so few coloured people in either country (and those who came rarely came to stay) that no general attitudes towards groups of coloured immigrants emerged until that time.

For the period during the two world wars, this meant that black Americans in the US armed forces gained a very favourable impression of Europe when they were stationed there. Afro-American writers, still second-class citizens in their own country fifty years after the Civil War that had purportedly been fought in large part over their position, were widely aware of the stance which the Europeans, with the French leading the way, had taken on slavery. In the First World War, many black Americans in particular volunteered to defend France: about 200,000 black Americans disembarked in the country in 1917. Whereas their white American commanders showed themselves supremely suspicious of their subordinates' relationships with white Europeans (particularly women), the French welcomed them and gave them a taste of life as full, equal citizens. W.E.B. DuBois concluded in 1919 that the black soldiers 'will ever love France'.[20] As Jennifer Keene commented in her detailed study of the subject:

> Because of African-Americans' positive experiences in France during WWI, Americans came to believe that France was a country without racial prejudice. A quick look at the experiences of West African troops calls this conventional conclusion into question. Both the Americans and the French were worried that the war would disrupt the racial status quo, and moved to limit the ability of their respective black populations to advance their civil rights during the war. The question still remains, however: why did the French treat African-Americans so well? The French had several stereotypes to choose from when meeting African-American soldiers. These stereotypes included their own view that West African soldiers were savage, child-like, and loyal; white American claims of African-American inferiority; the American view that France was a society without racial prejudice; and concerns about how America's wealth and power would affect France. Over time, the views of France as an egalitarian country and African-Americans as representing the best qualities of bravery, courage, and wealth overshadowed the others. In a large part, this transformation came about because of the paradoxical results of internal policing in the American Army. By heavily monitoring black soldiers, the American Army unintentionally turned these troops into ideal visitors in the eyes of French civilians. These civilians, in turn, enthusiastically accepted African-Americans' claim that France was a colour-blind society that (unlike the US) lived up to its ideals of fraternity, equality, and liberty.[21]

The situation was not that different in Britain, notwithstanding the British obsession with class which we have already commented upon. The British treated black GIs with a degree of respect to which they were by and large unaccustomed. General Eisenhower commented:

> The British population, except in large cities and among wealthy classes, lacks the racial consciousness which is so strong in the United States. The small-town British girl would go to a movie or dance with a Negro quite as readily as she would go with anyone else, a practice that our white soldiers could not understand. Brawls often resulted and our white soldiers were further bewildered when they found that the British press took a firm stand on the side of the Negro.[22]

There is anecdotal evidence from lower reaches of society to support this impression. A woman from Northern Ireland remembered a bus ride from her youth, in a wartime situation of 1942:

> [W]hen the Belfast bus stopped at the bus queue in Lisburn, several GIs boarded, some were coloured. When the bus was full, the conductor barred the door with his outstretched arm saying, 'Sorry, full up. ' A white American who couldn't get on stretched into the bus and drew out an unsuspecting coloured soldier. [The bus conductor] stepped out to block the white man's access and told the coloured man to get back into the bus and said to the white man 'No colour bar here, mate.'[23]

Sadly, racism spread to Britain and France just as it became more and more disputed in the US. This development produced a handful of critical films in Britain and France, which often, for good historical reasons, focus on cultural differences as much as on racism.

Black emancipation in film

The subjects of race relations and black emancipation of course have a long tradition in American film. While Second World War American propaganda films aimed at the US forces were still very careful to uphold segregation, blacks had long made inroads into American cultural elites through music, not only with jazz music, but also with more classical operas such as George Gershwin's *Porgy and Bess*, or musicals such as *Showboat*. Paul Robeson, the brilliant black lawyer-cum-marvellous bass singer, became famous not only for his main part in the 1936 film version of *Showboat* (he sang *Ol' Man River*), but also for his one-man civil rights campaign. A few notches down on the cultural scale, Harry Belafonte made an impression with the world-wide hit *Island in the Sun*. After the Second World War, Sidney Poitier did more than perhaps anybody else to dramatise anti-black sentiments in American society. Highlights in his campaign were *The Defiant Ones* (1958) in which he plays the escaped prisoner who has to get on with the redneck Tony Curtis, to whom he is physically chained; *The Heat of the Night* (1967), which exposed the enduring racism of the southern states; and *To Sir with Love* (1967) and *Guess Who's Coming to Dinner?* (1967), both of which showed up the double standards among American middle classes. Still, it took him until 2002 to be awarded an Oscar.

Other black actors have assumed the role of protagonist in Hollywood movies. Richard Rowntree was still a relative rarity when in 1971 he played a black detective in the American crime thriller *Shaft* (1971). Since then a series of comic actors have taken up the torch, especially Eddie Murphy with *Beverly Hills Cop I*, *II* and *III*, which in the 1980s became a genre in its own right. Will Smith in *Men in Black* walked in the footsteps of Eddie Murphy as cheeky, self-confident boy from the ghetto. Meanwhile, the *Bill Cosby Show* and later Oprah Winfrey regularly introduced black entertainers to American TV screens.

The subject matter of slavery and its legacy was tackled more seriously in Alex Haley's blockbuster novel *Roots* (1977), through Whoopy Goldberg's brilliant acting in Steven Spielberg's thoughtful film *The Color Purple* (1985), and then more humorously – incarnating common sense in a fight against bigotry – in *Sister Act* (1992). This topic, which had long been taboo in the US, has been brought out of the closet since the 1980s. The 1985 Merchant Ivory (significantly an Indian-British company) production *Jefferson in Paris* dealt with it in a film that portrayed the hypocrisy of Thomas Jefferson's support for the French Revolution, and his idealistic, never-consummated love affair with a white European, while at the same time depicting his sexual exploitation of his black slave, by whom he fathered a considerable number of descendants.

To measure inter-ethnic relations, one commonly applies the three markers of whether the groups in question live side by side (*cohabitatio*), actually mix, for example sharing feasts, celebrations, or attending common schools (*convivium*), or, finally, intermarry (*connubium*). Eating and feasting together (*convivium*), for example, applied to the menfolk of different religions and ethnic groups in the Ottoman Empire. Intermarriage (*connubium*) took place at a very substantial rate in Yugoslavia until the wars of the 1990s, showing that, even at that stage, the integration is reversible. Intermarriage (or love-relationships) between blacks (or native Americans) and whites to this day seems taboo in American films. If it does occur, one of the lovers usually dies tragically, no children born from the inter-racial relationships are shown, or if they are, they also play tragic roles and usually die.

This pattern is clear in Delmer Daves's films. In his *Broken Arrow* (1950), white James Stewart falls in love with Red Indian Debra Paget, and while the Indians condone their union, she is killed by white villains. In *The Last Wagon* (1960), Richard Widmark as the white hero was married to an Indian woman and had children with her (we are told), but all three (whom we do not see) have been killed by whites. He encounters a mestizo girl who is the most positive among the 'white' characters, but who is duly shot. In *Dances with Wolves* (1990), Kevin Costner, himself acculturated by the Indians, ends up in the film with a mate, who this time is genetically 'white' (the big concession to multi-ethnicity being that both have adopted Indian culture). It takes a foreign director with *The Day after Tomorrow* (2004) to show a mixed (black and white) couple with a child that actually survives the catastrophe! A rare exception to the taboo of mixed relationships is the already mentioned *Guess Who's Coming to Dinner*, in which the white middle-class girl brings home her black fiancé (Sidney Poitier) to meet the (shocked) parents, played convincingly by Katherine Hepburn and Spencer Tracy. But the taboo is still upheld almost thirty years later in the 1996 film *The Long Kiss Goodnight*, in which a white heroine (Geena Davis) teams up with a black partner (Samuel L. Jackson) in investigating a

CIA-organised assassination programme: the absence of any erotic relations between them is assured by the age difference between the two characters.

The emancipation of coloured peoples in the film industry can also be mapped in the Hollywood-produced James Bond films, which make concessions to American susceptibilities about race: Japanese ladies become acceptable as lovers first. In *Dr No* (1962), a Japanese lover turns out to be a traitor and duly succumbs to the bite of a poisonous snake. In *You Only Live Twice* (1967), James Bond, himself posturing as a Japanese, has two successive Japanese lovers, both of whom have positive roles: the first is poisoned in an attempt on Bond's life and dies for him. In *Tomorrow Never Dies* (1997), he teams up with Chinese Michelle Yeo. The first black woman with whom he grapples is wicked (Grace Jones in *A View To Kill*, 1985). Only in 2002, in *Die Another Day*, does he have a black lover (Halle Berry), who is a 'goody', and – surprise, surprise – a 'Cousin', that is, an American secret agent.

All of this is not to deny that Britain and France have their own problems with racism and cultural differences. But in both societies the *connubium* seems more acceptable (at least on screen) than on American celluloid. The social problems this poses in a conservative society are not dodged, but usually faced head-on. In British novels and films, the central subject tends to revolve around Anglo-Indian relations, for example in the novels, and subsequent film versions of, E.M. Forster's *Passage to India* and *Heat and Dust* (which revolves around the scandalous elopement of a British colonial official's wife with an Indian prince), and the *Jewel in the Crown* trilogy of novels and its Granada television version, in which one central story is the love affair between a British-educated Indian and a white Englishwoman, who dies at the birth of what is either their child or a child engendered when she was gang-raped by Indian anti-colonial rebels.

Perhaps Martin Lawrence as the Black Knight marks the completion of the emancipation of American blacks on celluloid? He is in any case portrayed as attractive to white women (which would have been taboo fifty years earlier). Pursued by an English (white) princess, he is tricked into a sexual relationship under false pretences. It is probably a sign that this is no longer thought scandalous that it is not presented as unusual. But it is also the case that Jamal himself courts a *coloured* heroine, with whom he manages to secure a 'date' at the end of the film. The coloured heroine Jamal courts miraculously follows him from his medieval English adventure to the present. The riddle of quite how is not resolved. Perhaps she is a reincarnation rather than the same person, as she does not seem to remember having met Jamal before. Before he can find out, another bad fall leaves him unconscious again, this time transported into the middle of a Roman arena, lions about to pounce on him. (Fear not, the 'Peplum' film will not be discussed in this chapter.)

Memories of British gallantry or oblivion of British loyalty?

Let us return to the interaction between Old World and New World. Jamal upon his return from the fictional medieval 'England' is not who he used to be. As in any good *Bildungsroman*, he has undergone a moral evolution, maturing, awakening. Finding the lady of his heart and defending her in an oppressive hierarchical society (in which she was molested by a knight not on account of her skin colour but on account of her slightly lower social standing in this medieval, feudal society) was one part of this evolution. Jamal suddenly turned chivalric himself. Moreover, from the Robin Hood-type resistance of the rebels against the unlawful (and lawless) king, Jamal takes away an idealism which replaces the petty materialism he displayed at the beginning of the film. The idealism he acquires concerns social responsibility, which for him now ranks higher than his personal gains. In his journey to the Middle Ages, he has joined in the fight for freedom and against tyranny. Indeed, it is his dangerous and potentially self-sacrificing action which has led to the victory of the rebels against the tyrant.

Jamal is an echo not only of Robin Hood, but also of the role in which Britain has been cast by the US, particularly in that greatest fight of Good against Evil in human history, the Second World War. The eighteen months in which Britain 'stood alone' in the face of a genocidal tyranny that controlled almost all of Europe turned the country as a whole into the world's bastion of freedom against dictatorship. This image was carefully cultivated even before America's entry into the war in December 1941 in US films sympathising with Britain, the greatest classics being *Mrs Miniver* and *Correspondent 17*. All American films about the Second World War made since tend to portray the British as 'gallant allies' and fairly admirable, self-sacrificing defenders of the West; this even applies to Disney movies, memorably in *Bedknobs and Broomsticks* (1971), in which a middle-aged British spinster-turned-witch conjures up the ghosts and armours of Agincourt and Blenheim to scare off a German invasion of Britain. While this subject is not explicitly alluded to in *Black Knight*, British gallantry strongly resonates in the film.

However much or little Britain is remembered on American celluloid as gallant ally in the Second World War and Iraq, the impression that one comes away with in our literary-cum-cinematographic stroll is that, in fact, it is not Britain that Americans on the whole perceive as progressive. Instead, it is (politically stroppy) France that is America's political twin and rival at one and the same time: an alternative New World to that incarnated by the United States. France embodies the spirit of a 'New Europe', Britain on the whole the spirit of an 'Old Europe'. And yet, Britain is seen as the cradle of a lineage of rebellion and a quest for freedom from the tyranny which led directly to the American War of Independence. In this way, Britain is remembered as the (old) ancestor of America.

At all times, Britain has differing aspects. There is, as we have seen, England, which stands both for royal tyranny, Anglo-Saxon religious intolerance, imperialism and class oppression, and for the rebel lineage from Robin Hood and Wat Tyler to American liberty. There is the Celtic heritage, which although English-speaking is decidedly anti-English. These are some of the facets of Britain's depictions in American films, and clearly, as we have seen, contradictory facets. A 'special relationship' between Britain and the US is documented by this, as there are probably more cultural references to Britain, British cultures (in the plural) and British history in American cinema than references to any other 'foreign' culture, and probably more British actors in Hollywood than other foreign-born actors.

The American film industry in the service of European cultural heritage

This raises a number of fascinating questions with regard to the role of popular culture – or cultures, as one should say – in a globalised world. Outsiders so often speak in simplistic terms of American popular culture which is set on conquering the world, of Hollywood imperialism, driven by American economic interests. For one, many of the examples mentioned above illustrate that American movies transport more than one culture, or rather, transport a great variety of cultural symbols and references. Many of these symbols and references that have their roots in British culture(s) and history in turn are intelligible to many English-speaking cultures of the world – from India to Australia, from Canada to New Zealand. Many other cultures have become familiar with essentials of British culture via Hollywood, from Robin Hood via *Shakespeare in Love* to Peter Pan. The same, on a slightly lesser scale, is true for French culture: it is probably due to the Hollywood products rather than to the wide reading of Alexandre Dumas in translation that Musqueteer costumes for children to be worn at *fiestas* are on sale in shops around Spain (indeed, made in Spain). American films thus transport cultural signifiers that have their origins outside America, in some cases bringing them back to their countries of origin, or to cultures with roots in those countries of origin, or to cultures generally encountering them for the first time. The cinema of the New World thus globalises popular cultural heirlooms of Europe, and that definitely contributes to its richness, its popularity, and to the receptiveness of many parts of the rest of the globe that are at least in part cultural descendants of Europe. Now as ever, America is thus spreading European culture, European ideas, a European popular heritage. European culture has long exploited Hollywood for its own transportation, one might say, an answer worth making to many European laments about American cultural imperialism. It raises the question as to where American culture ends and European cultures begin, and as to whether one should be talking about a

Western culture more generally rather than about distinctive American and European cultures. We thus have to recognise the coexistence of many cultures, and of many levels of culture, that can coexist in deepest contradiction. American cinema can thus transport anti-English feelings while globalising English history and popular culture. It can be read as 'Western' culture as well as Scottish nostalgia. Hollywood films tap into European myths and fairytales as well as legends and historic events well-packaged in literary traditions that are already faintly known, vaguely familiar, and usually associated with olden times. The New World popularises images of Old Europe in a global society.

Thus American screen-images can be read in different ways. They can appeal to Anglophobia and to Anglophilia (or Francophilia, respectively), and at the same time they can kindle nostalgic interests in heroic episodes of (vaguely remembered, and therefore common) history. If anything, these multi-layered images and the many messages they can transport are a testimony to the complexity of the relations between imagined communities, which in centuries of coexistence have formed manifold (and often contradictory) images not only of themselves, but also of other communities, be they friendly, rival or at times antagonistic. Any reduction to just one image or metaphor is as evocative of past myths as it is puzzling to those more familiar with the complexity of these relations, which have grown and shifted backwards and forwards over time, and which have a rich heritage of literature, and more recently film, to draw upon. This amply confirms the views of Shimko, Burgess and Steinbruner quoted at the beginning: it underscores both the longevity and the mutability of images, belief patterns and perceptions.

Notes

1 Keith L. Shimko, *Images and Arms Control: Perceptions of the Soviet Union in the Reagan Administration* (Ann Arbor: University of Michigan Press, 1991), pp. 11ff and 25.
2 P. Burgess, *Elite Images and Foreign Policy Outcomes* (Columbus: Ohio State University Press, 1968), pp. 4, 7.
3 John Steinbruner, *The Cybernetic Theory of Decision* (Princeton: Princeton University Press, 1974), p. 130.
4 Beatrice Heuser, *Nuclear Mentalities* (Basingstoke: Macmillan, 1998).
5 Jeane Kirkpatrick, *'We and They': Understanding Ourselves and Our Adversary* (Washington DC: Ethics and Public Policy Center, 1983).
6 Smith Burnham, *Our Beginnings in Europe and America: How Civilization Grew in the Old World and Came to the New* (Philadelphia: John C. Winston, 1918), p. 185.
7 Ibid.
8 John Dumbrell, *A Special Relationship: Anglo-American Relations in the Cold War and After* (Basingstoke: Macmillan, 2001), pp. 196–219.
9 Philip Rahv (ed.) *Discovery of Europe: The Story of American Experience in the Old World* (Garden City, NY: Anchor Books, Doubleday, 1960), p. 4f.

10 Ibid., p. 8.
11 Zitiert in Juliet Gardiner, *'Over Here': The GIs in Wartime Britain* (London: Collins & Brown, 1992), p. 59.
12 Norman Hampson, *The Perfidy of Albia: French Perceptions of England during the French Revolution* (London: Macmillan, 1998), pp. 61–77.
13 George Bancroft, *A History of the United States from the Discovery of the American Continent to the Present Time*, 6 vols (New York: D. Appleton & Co., 1834–85).
14 Sarah J. Purcell, 'Lafayette, Memory, and American Democracy', in William L. Chew III (ed.) *National Stereotypes in Perspective: Americans in France, Frenchmen in America* (Amsterdam: Rodopi, 2001), pp. 67–88.
15 Rahv, *Discovery of Europe*, p. 57f.
16 Quoted in David Strauss, *Menace in the West: The Rise of French Anti-Americanism in Modern Times* (Westport, Conn.: Greenwood Press, 1978), p. 15.
17 Larisse Hilbig, *Der Mythos Paris in der amerikanischen Literatur* (Frankfurt/Main: Peter Lang, 2003).
18 Quoted in André Wilmots, *Le Défi Français: ou, La France vue par l'Amérique* (Paris: Bourin, 1991), p. 15.
19 Jean Guisnel, *Les pires amis du monde: les relations franco-américaines à la fin du XXe siècle* (Paris: Eds. Stock, 1999).
20 W.E.B. DuBois, 'The Black Man in the Revolution of 1914–1918', quoted in Tylar Stovall, *Paris Noir: African Americans in the City of Light* (Boston: Houghton Mifflin, 1996). See also Michel Fabre, *From Harlem to Paris: Black American Writers in France, 1840–1980* (Urbana, Ill.: University of Illinois Press, 1993).
21 Jennifer D. Keene, 'French and American Racial Stereotypes during the First World War', in Chew, *National Stereotypes*, pp. 261–81.
22 Zitiert in Juliet Gardiner, *'Over Here'*, p. 155.
23 Ibid., p. 154.

14

'DIVIDED WE STAND'

America's rhetoric of defence and defensive European politics

Walter W. Hölbling

What on the American side is seen as justified defensive rhetoric in the face of an imminent external threat often, though not always, comes across as an aggressive-defensive attitude when viewed from the outside. Consider, for example, the following quotations, separated by over three centuries. The first is from a contemporary historical account of the war between the Puritan settlers in New England and the Native Americans, and the other two are from recent speeches given by the present American President.

> That the Heathen People amongst whom we live, and whose land the Lord God of our Fathers hath given to us for a rightful Possession have [. . .] been planning mischievous devices against that part of the English Israel which is seated in these goings down of the sun, no Man that is an Inhabitant of any considerable standing, can be ignorant.[1]

> Iraq continues to flaunt its hostility toward America and to support terror. The Iraqi regime has plotted to develop anthrax, and nerve gas, and nuclear weapons for over a decade. [. . .] States like these, and their terrorist allies, constitute an axis of evil, arming to threaten the peace of the world. By seeking weapons of mass destruction, these regimes pose a grave and growing danger. They could provide these arms to terrorists, giving them the means to match their hatred. They could attack our allies or attempt to blackmail the United States. In any of these cases, the price of indifference would be catastrophic.[2]

> There are still governments that sponsor and harbor terrorists – but their number has declined. There are still regimes seeking weapons

> of mass destruction – but no longer without attention and without consequence. Our country is still the target of terrorists who want to kill many, and intimidate us all, and we will stay on the offensive against them, until the fight is won.[3]

Richard Hofstadter traced this particular aspect of American rhetoric in his book, *The Paranoid Style in American Politics* (1966), and he found it characterized by the tendency to secularize a religiously derived view of the world, to deal with political issues in Christian images, and to colour them with the dark symbolism of a certain side of Christian tradition. Social issues could be reduced rather simply to a battle between a Good and an Evil influence.[4]

There is ample evidence that variations of the original Puritan version of the 'errand into the wilderness' have strongly coloured official as well as popular American discourse in times of catastrophe, or fear of catastrophe, or in fictional scenarios of catastrophe. Over the years, these threats have had many faces, most of them external – wild nature, Indians, other religions and ideologies (Jacobins, Freemasons, Catholics), and foreign powers – from the Spanish and the French and the King of England to the Mexicans, Habsburgs, the German 'Huns' in World War I, the Axis powers in World War II, World Communism with the USSR and China, and its outposts in Cuba, Vietnam, Chile, Nicaragua and Grenada, and most recently the so-called 'rogue states' and international terrorism. At various periods there were also immigrants from 'exotic' countries (remember the restrictive Immigration Acts of the 1920s), cyber-terrorists, foreign drug-lords, and aliens from outer space. Some of these threats have also come from within – among them African-Americans, emancipated women, leftish unions, Hollywood, conspiring power-hungry politicians, entrepreneurs and scientists, alcohol, organized crime, communist spies, and smokers, as well as supporters of dangerous movements like pro-choice, multi-lingualism, and same-sex marriage.

In short, seen from a European perspective, the unifying American master narrative seems to be in constant need of an identifiable threat, preferably from the outside, in order to maintain its persuasive power and to provide the necessary cohesion among increasingly differentiated political/cultural/ethnic interest groups within its dominion. The most recent examples of this US rhetoric of defence followed the terrorist attacks on the World Trade Center of September 11, 2001, and also prepared the ground for the US invasion of Iraq in the spring of 2003. The attacks of 9/11, in the eyes of Europeans as well as most other nations, justified a firm statement of defence and made even old antagonists take the side of the US; in fact, support for the US was practically global, even though the comprehensive coinage 'war on terrorism' caused misgivings among some who saw in this phrase the seed of a never-ending mission conducted under rather arbitrary criteria.

Very differently, the American decision in 2003 to invade Iraq was only underwritten by a rather numerically challenged 'coalition of the willing' whose actual construction belied the White House claims of wide support. Within one year, global support for the US had dwindled dramatically, and in Europe was limited to those nations who were traditional allies (e.g. Great Britain), those whose governments expected to increase their regional political clout (e.g. Italy, Spain), or those who were 'persuaded' by the threat of the withdrawal of US military or economic funding (e.g. Bulgaria, Romania). The US decision to start the war against Iraq was seen as a gross example of American unilateralism that undermined the United Nations and simply practised the maxim 'might makes right'. The flip-flopping arguments for the necessity to go to war that kept pouring from the White House, supported by little more than shaky 'evidence', were not helpful either and convinced only a few: Iraq's supposed threat to the world because of its stock of weapons of mass destruction, Saddam Hussein's alleged connection to al-Qaeda, the liberation of Iraqi citizens from dictatorship, and the establishment of democracy – all of these were considered to be rather rhetorical allegations, if not outright fabrications, by most Europeans. Two years later, in 2005, no weapons of mass destruction had been found; no proof for a connection between Saddam Hussein and al-Qaeda had been made public; the dictatorship in Iraq had been overthrown all right, yet little more than a formal democracy had been established, and its development was threatened by ongoing terrorist attacks that had cost thousands of lives since the official 'end' of the Iraq war; and the daily news reports of the dead and wounded had become painfully reminiscent of the notorious 'body counts' during the Vietnam conflict.

The United Nations as well as those European leaders who refused to go to war not only have had to defend themselves against a flood of attacks from US politicians and the US media; they are now asked to help the US to clean up the chaos the 'coalition of the willing' has caused in Iraq, and they also have to face some long-term effects of the split the war caused among members of the European Union. Intentionally or not, the US invasion of Iraq has so far destabilized the Middle East, has turned Iraq (formerly a secular dictatorship) into a playground of assorted fundamentalist terrorist groups and antagonistic religious and political factions, and has caused growing dissent among the countries of the European Union in a highly strenuous phase of expansion (ten new members) and simultaneous consolidation (the proposed European Constitution). It is therefore no surprise that Europeans have become a bit wary of their partner across the ocean, and even though few seriously question either the existence of or the need for some transatlantic community, they have begun to look at its elements with a more critical eye, sobered by recent events.

Since the time of William Bradford's *History of Plymouth Plantation* and Mary Rowlandson's paradigmatic first Indian captivity narrative – the story of a woman of European descent kidnapped by Native Americans – when nature and indigenous inhabitants of the American continent still constituted real dangers for the small numbers of European colonizers, there has developed what one might call an *asynchronicity* between the actual threats and the ones identified in the dominant discourse. In other words, the rhetorical figures and images employed in this discourse are often out of synch with historical realities. This implies, among other things, that:

- Survival and acceptance of traditional rhetorical and narrative concepts and images continue much longer than the actual historical situation would warrant. These constructs can be considered examples of a 'storifying of experience', as they employ specific symbol systems, myths, narrative structures and modes of discourse that are considered as adequate conceptual frames for the understanding of a historical situation. If they have sufficient explanatory power, these models of rhetorical sense-making persist as conventions even in the face of political and historical inadequacy.
- Because of their familiarity they can easily be instrumentalized for political, religious and economic purposes.
- Even in the face of contradictory factual evidence, they can serve as formulaic rituals in times of crisis, uniting the nation against a real or imagined danger.
- In fact, the usage of these words sometimes tends to gain a life of its own which may actually create the situation it supposedly tries to avoid.

A very obvious example is the American 'Indians' as the ubiquitous enemies in US Western novels and later on in Hollywood movies at times when, historically speaking, Native Americans had not been a real danger to the development of the nation for several generations. Similarly persistent, though different in its origins, is the image of the potent black male lusting after white maidens, in spite of the proven fact that the historical reality of the situation has been rather different. On another level, since the 1980s we can observe an 'English only' movement in several US states, which tries to legally guarantee the use of English as the only official language. From a European perspective, this attempt is as incomprehensible as it is unnecessary: as if one could 'protect' a language, in the first place; as if English (of all languages!) needed any protection; as if knowing/speaking more than one language would make you less 'American'; as if the belief in common American values were dependent on one particular language.

As it is, as early as 1889 Mark Twain's satirical novel *A Connecticut Yankee in King Arthur's Court* radically questioned the American stereotype of the lone cultivator in the wilderness and his mission. The text clearly shows structural characteristics of the 'captivity narratives', but the traditional situation of the captive has been turned upside down. Here, it is not the representative of the civilized world who is permanently threatened with death and/or moral degradation by his transfer into a 'primitive' society. Rather, it takes Twain's hero Hank Morgan, foreman of an arms factory, only a few years to uproot the social and spiritual order of King Arthur's medieval England and threaten it with extinction.[5] That he cannot succeed in the end is due to the logic of the story but does not invalidate Twain's critique which, one should remember, was voiced at a time when Native Americans in the US came as close to extinction as never before or after.

While deconstructing, on the one hand, an outdated American stereotype, Twain also presents us with an exaggeratedly drawn alternative stereotype according to which Americans, seen from a European point of view, excel in practical ingenuity and business skills, but are rather naive, provincial and underdeveloped as regards cultural sensibility and creativity. The Yankee – with the best of intentions, mind you – as destroyer of a European medieval culture whose values and spirituality remain alien to him, comes across as a glossy pop-art version of those fictional American travellers in Europe that populated the novels of Henry James at the time (*The American*, 1877; *Daisy Miller*, 1878; *Portrait of a Lady*, 1880/1), who try to re-assess, usually with moderate results, their relationship to the cultures on the other continent. It is also a theme which, as Beatrice Heuser points out, was taken up by Hollywood in a number of versions of Twain's novel (see Chapter 13).

Finally, Twain's *Connecticut Yankee*, with wonderfully ambiguous irony, not only reverses the conventional structure of the captivity narrative, it also expands it and makes it international: now the wilderness to be cultivated is Old England, and the Indian braves there are the ancestors of the New World cultivators. Twain's satire also implies that the new American interest in their European origins carries the seed of an expansionist re-conquest, given the fact that the American continent has been officially settled and new frontiers must be sought outside the continental US. Not surprisingly, Twain's novel does not seriously impair the validity of the national stereotype of the righteous American hero defending Faith and Civilization against the onslaught of barbarian hordes. In World War I, this lent itself easily for use in the American view of the situation in Europe: Americans as 'Knights of Democracy' in the 'Great Crusade' against 'the Hun', in order to save 'La Belle France', 'Innocent Belgium' and 'Classical Italy' (symbolizing European culture) from destruction. This conceptual framework projects the symbol system of the 'captivity narratives' onto the international scale, complete with all the major components of missionary zeal, racial warfare, gender-specific roles of victim and saviour, and their not so implicit sexual connotations.[6] With much less public rhetoric, the country sending its soldiers across the Atlantic to 'make the world safe for democracy' over the period from 1898 to 1925 took advantage of a Europe torn by war, and practised an expanded version of the Monroe doctrine among its southern and Pacific neighbours with military involvements in Panama (1908, 1912, 1925), Nicaragua (1909, 1910, 1912–25, 1926), Cuba (1898, 1906, 1912, 1917–23, 1933), Mexico (1914, 1916–17), Haiti (occupied 1915–34), the Dominican Republic (1904: financial system; occupied 1916–24), Honduras (intervention 1907), Guatemala (1921 coup against president), and Puerto Rico (occupied 1898), as well as Guam (1898) and the Philippines (1898).

One might argue that World War II was one of the few situations after the American War of Independence in which the dominant US discourse of external danger corresponded to a real historical threat from external enemies. Beyond question there was a broad consensus that resistance against the Nazis and fascists in Europe and their allied Japanese imperialists in the Pacific was justified on political and moral grounds. And although in its aftermath a few critical American authors like Norman Mailer, Irwin Shaw, John Hersey and others, while supporting the war goals, pointed to its potentially dangerous effects for the victors, their voices remain an influential minority. The dominant discourse tells the kind of story that follows the traditional master narrative, as Ward Just ironically sums it up in his study *Military Men* (1970):

> Since American wars are never undertaken for imperialist gain (myth one), American soldiers always fight in a virtuous cause (myth two) for a just and goalless peace (myth three). [. . .] American wars are always defensive wars, undertaken slowly and reluctantly, the country a righteous giant finally goaded beyond endurance by foreign adventurers.[7]

Historically, the outcome of World War II vindicated this self-image to some degree, but soon the escalation of the Cold War darkened the picture, internationally as well as domestically. US foreign policy at the time included the standard ingredients of exceptionalism, moral superiority, democratic mission, and The Enemy, now 'World Communism'.

In 1966, Senator J. William Fulbright, then Chair of the Senate Foreign Relations Committee, published his critical assessment of the US government's practice of seeing international aid programmes as an instrument for maintaining an 'American presence' and for spreading the 'Great Society'. He specifically mentioned the then escalating Vietnam conflict and commented:

> These [aid] programs are too small to have much effect on economic development but big enough to involve the United States in the

> affairs of the countries concerned. The underlying assumption of these programs is that the presence of some American aid officials is a blessing which no developing country, except for the benighted communist ones, should be denied.
>
> I think this view of aid is a manifestation of the arrogance of power. Its basis, if not messianism, is certainly egotism.[8]

In the same year that Fulbright published his critique of US foreign policy, the European mutant of a classical American movie genre became an instant box office success in the US (as was its European release one year before): Sergio Leone's *The Good, the Bad and the Ugly*. The film apparently belongs to one of the most 'American' genres but leaves few of the key elements of the traditional American Western unturned. The plot somewhat arbitrarily develops the chase of three gunmen after a batch of Confederate gold during the last years of the American Civil War. Individual exceptionality still plays a role but is no longer permanent or absolute. Moral superiority is virtually absent from the movie: the driving forces motivating the protagonists' actions are greed and power, with occasional sadism thrown in for 'emotion'. A democratic mission is nowhere in sight, neither by the few representatives of a civil society nor by the government institution in the movie, the Union Army, whose only two honest representatives are obviously helpless against corruption in the ranks as well as against an incompetent higher command. The Enemy is practically everywhere – in other words everyone who competes in the race for money and power. The Good (Clint Eastwood) has the upper hand when the movie ends, but we know that his streak of luck can end any time. There are no real heroes: The Good is only 'good' compared to the calculated viciousness of The Bad (Lee van Cleef) and the mindless thuggery of The Ugly (Eli Wallach). The film projects a world torn by war; law and order are either inefficient or virtually absent, corruption is rampant, and individual survival depends on a loaded gun and the whims of fortune.

In short, *The Good, the Bad, and the Ugly*, while claiming affinity to the genre of Western movies, radically deconstructs it; what we have is a morality play without morals. Although there are no overt political overtones, the film presents a world that must have appeared familiar to its contemporary audiences in the US as well as in Europe in the mid-1960s. At a time when consensus in US society was violently threatened by apparently unbridgeable differences, Leone's scenes of senseless death and destruction in the Civil War not only reminded American audiences of the – until then – most painful period of their national history; it also called to mind the ongoing war in Vietnam and allowed for contemporary connotations. The identification patterns available in the movie are multiple and opaque enough to even be contradictory.

For example, the battle scenes as well as the corruption and sadistic abuse of power portrayed in the movie speak strongly to those protesting against government authority, the selective draft, and the war in Vietnam. Because of the allegorical quality of the protagonists and the mythical story line, though, the movie also enables much more intriguing readings: US Southerners can see 'The Bad' as a representative of the Yankee North, and his death as poetic justice in the context of the American Civil War. However, supporters of the Vietnam conflict – remember, this was officially a civil war between South and North Vietnam in which the South had appealed to the US for help – also can easily allegorize 'The Bad' as the cruel North Vietnamese (Ho Chi Minh), and 'The Ugly' as the corrupt South Vietnamese regime which is repeatedly saved by US intervention ('The Good') just in the nick of time before its demise. In fact, in the episode where The Good ends his contract with 'The Ugly' and leaves him out in the desert, contemporaries might establish parallels to the 1963 'removal' of Ngo Dinh Diem by a military coup, with the quiet cooperation of the US. On yet a different level, critical viewers can understand the shifting alliances between the movie's protagonists in their pursuit of the booty as suggestive of US foreign policy since the end of World War II, which started off with a major reversal of the war-time alliance and continued to make liaisons of convenience according to the necessities of *realpolitik*.

There is also a more general element in Leone's film: it is, as today's anti-globalization protesters would argue, the disturbing fact that individual existence is rather precariously at the mercy of global economic and power games, and that wars in this system have nothing to do with emotions, morality, religion, or ideology. This aspect is especially of concern to a society like the US, where the discourses of individual rights and freedom as well as of moral/religious obligation have been the major pillars of the national cultural fabric, as well as a dominant rhetorical element in the justification of all wars the US has ever fought. The emotionally and morally arid world of Sergio Leone's movie, which in effect suggests that killing is good business, has no place for these lofty rhetorical aspirations and, in fact, corresponds quite well to Joseph Heller's terse and darkly suggestive alliteration a few years earlier in his novel *Catch-22*: 'Business boomed on every battle front.'

A response of a different kind to the American situation in the 1960s is Norman Mailer's 1967 novel *Why Are We in Vietnam?* His answer to the title is a kind of fictional psychoanalysis of the collective American unconscious, illustrated by the story of a high-tech hunting party of Texan corporate executives in Alaska. But Mailer goes far beyond suggesting easy analogies between this hunting trip and Vietnam; he is looking for the roots of this unbridled joy of killing, of the fascination with high-tech overkill in the collective American psyche. He articulates his belief that at the bottom of it all are the accumulative and mutually reinforcing effects of repressive sexual norms, secularized versions of the Puritan work ethic, business interests and the military-industrial complex, American imperialism backed by an unbroken sense of mission, the belief in 'manifest destiny', and a holy fear of everything that does not conform to the WASP way of life – including the notorious suppression and commodification of the body, human or animal.

Mailer's most irreverent indictment of this American attitude appears in the middle of the novel when Rusty Jethroe, the leading CEO, having failed to prove himself as the top big-game hunter in front of his subalterns, is distressed and ruminates about the possible consequences of this embarrassing situation:

> Yeah, sighs Rusty, the twentieth century is breaking up the ball game, and Rusty thinks large common thoughts such as these: 1 – The women are free. They fuck too many to believe one can do the job. 2 – The Niggers are free, and the dues they got to be paid are no Texan virgin's delight. 3 – The Niggers and women are fucking each other. 4 – The yellow races are breaking loose. 5 – Africa is breaking loose. 6 – The adolescents are breaking loose including his own son. 7 – The European nations hate America's guts. 8 – The products are no fucking good any more. 9 – Communism is a system guaranteed to collect dues from all losers. 9a – More losers than winners. 9b – and out: Communism is going to defeat capitalism unless promptly destroyed. [. . .] 12 – The great white athlete is being superseded by the great black athlete. 13 – The Jews run the Eastern wing of the Democratic party. 14 – Karate, a Jap sport, is now prerequisite to good street fighting. [. . .] 17 – He, Rusty, is fucked unless he gets that bear, for if he don't, white men are fucked more and they can take no more. Rusty's secret is that he sees himself as one of the pillars of the firmament, yeah, man – he reads the world's doom in his own fuckup. If he is less great than God intended him to be, then America is in Trouble. They don't breed Texans for nothing.

While Increase Mather's self-image in the quotation at the beginning of this chapter can still be validated by religious beliefs and the historical situation, Rusty's view of himself and the endangered state of White America comes across as the compensating aggressiveness of a power elite that tries to hide the lack of an ethical and ideal core for their claim of supremacy behind grandiose appeals to America's God-given greatness.

Mailer's 1967 critique has gained almost uncanny topicality. Recall that on March 8, 1983, speaking before church leaders in Florida, then US President Ronald Reagan named the Soviet Union as the seat of 'evil in the world'. With the collapse of the Soviet empire in 1989, the demonizing

Cold War rhetoric temporarily disappeared from US public diction, yet it immediately returned in the wake of the terrorist attacks of 9/11. For his 2002 'State of the Union Address' on January 29 of that year, President George W. Bush's speech writers coined the term 'axis of evil'. Apart from the fact that the two terms are 'four-letter words', it is a clever choice of phrase that evokes the Axis powers of World War II as well as the Cold War, and also suggests an American moral superiority of the fundamentalist kind. This rhetoric places the US once again on the side of God in a primeval show-down against the forces of darkness in which American soldiers' bodies – and their electronic and high-tech extensions – are the primary weapons.

In this 2002 State of the Union address, the combination of 'war against terrorism' and 'homeland security', with the envisioned beneficiary effects of 'safer neighbourhoods' resulting from 'the sacrifice of soldiers, the fierce brotherhood of firefighters [. . .], stronger police [. . .] stricter border enforcement' and America's dependence on 'the eyes and ears of alert citizens', convey connotations which for many European ears have an ominous ring. This rhetoric can be situated in the context of recent American events: major business malpractice (such as the Enron, WorldCom and NYSE scandals), neglect of environmental concerns (the unsigned Kyoto agreement, oil projects in Alaskan natural reserves, cutting of federal detoxification funding), blatant violations of civil liberties in connection with people detained indefinitely without legal assistance following 9/11, the creation of the Homeland Security Agency, the Patriot Acts, the decision to dramatically upgrade military weapon systems, and the revelations in testimonies before the Congressional 9/11 Commission, as well as US and British intelligence agencies' reports on the actual state of information in regard to Iraq's weapons of mass destruction.

To make it clear – this is not to comment on the *pragmatic effectiveness* of policies adopted by the US administration, nor to draw superficial analogies. What is worth contemplating, however, is the structural and thematic affinities to Mailer's satirical fictional analysis of US society in 1967, then in another state of crisis. They seem to imply – and some people might find *this* a bit alarming – that conceptual changes in the mind of American leadership over the past thirty-eight years have not been very significant. The 'Domino Theory' and the 'weapons of mass destruction' share a certain ringing rhetoric. Intentionally or not, the adamant religious and moralizing diction in the statements of the current US administration bears strong affinities to the venerable biblical rite of the scapegoat. James Aho in his 1981 study on *Religious Mythology and the Art of War* says the following:

> As a rule, in Judaism, Islam, and Protestantism, responsibility for the world's sin is projected onto minority populations, strangers, and foreigners; those with tongues, customs, and pantheons alien

> to God's faithful. In collectively objectifying evil and positing it upon this external enemy, a sense of cleanliness of His 'remnant' is created symbolically. Analogous to the Levitical rite of the scapegoat (Lev. 16: 20–12), the projectors can 'escape' from acknowledging the possibility of their own blemish. [. . .] Thus, mythologically, the holy war will be fought between the absolutely righteous and the equally absolute incarnation of Evil. Insofar as it exorcises the objectified evil, the ferocity of the violence in the war must reflect the enormity of the crime against God and man. [. . .] The Hebraic, the Muslim and Christian holy wars, both in myth and enactment, are among the most ruthless in human experience.[9]

The current structure of the George W. Bush–Osama bin Laden–Saddam Hussein scenario, the 'coalition of the willing' versus 'the axis of evil', 'rogue states', and global terrorism of the Islamic fundamentalist kind, looks very much like the biblical scapegoat ritual. It conveniently allows *both* sides to 'escape' from acknowledging the possibility of their own blemish by creating a Manichean system of absolute good versus absolute evil. This also seems to have become an increasingly accepted view in pop culture products: for example, *Lord of the Rings* and similar recent box office hits. Or, a rather quotidian – almost banal – example to illustrate the contemporary 'official spirit' in the US: in January 2005 a display of the four latest special editions of stamps included one against breast cancer. The other three featured John Wayne, a Purple Heart, and the National War Memorial. If this is an indication of how the contemporary American psyche may be moulded, it brings to mind Golda Meier's alleged observation to Henry Kissinger: 'even paranoids have enemies'.

On the US side, the somewhat arbitrary shift in scapegoats – from the elusive Osama bin Laden and his al-Qaeda network to the more targetable Saddam Hussein – also provides the latest example of the aforementioned asynchronicity of the discourse, whatever political, economic and strategic motivations might be behind that move. The actual *new* historical causes of the contemporary danger – Islamic fundamentalist terrorism growing out of poor social, political and economic conditions – are ignored in favour of a familiar threat (Iraq) that can be easily identified and attacked, though in historical reality it has not been a real global danger since 1991. Given the current situation, it is likely that the attitude of Europeans towards America will see yet another turn of the critical screw, even though the recent diplomatic and charm offensive towards Europe and a more decided US attitude in the Near East seems to indicate that the US is interested in refurbishing its image in these regions. Yet European scholars might do well to make greater analytical efforts to understand what on the surface come across as irreconcilable American opposites, for example, fundamentalist religious beliefs and a free democratic system, or the claim that in elections 'every vote counts', though the actual voting/counting of votes (mechanical or electronic) is subject to procedures that leave many Europeans stunned. More attention might also be paid to the extremely mediated and visual quality of everyday US life, as well as to the impact this has on popular understanding of democracy and its processes. Thanks to the rhetoric of the US administration in the lead up to the Iraq war, 50 per cent of Americans believe that Osama bin Laden and Saddam Hussein actually cooperated, though serious evidence for that has not become public yet.

America's claim of exceptionalism, and for leadership in the democratic world, increasingly has to defend itself against the charge that there is nothing special about the US: that it, like any other imperial power in history, uses military force whenever necessary to secure its interests, and it only cooperates with the international community when it is expedient for American interests to do so. In short, a re-assessment of the role of America in a post-Cold War world will centre on whether the US can convincingly act as the leader of democratic nations, or rather come across as the global bully. What US Secretary of Defence Donald Rumsfeld called, in 2003 (rather mistakenly), the 'Old Europe' – in his terms those nations who did not follow the US into the war – constitutes, in fact, the 'New Europe', a Europe that after centuries of devastating wars has finally decided that military force is not the first but the last instrument to decide differences among nations. From a European point of view, it is a very worrying counter-development that the US, until World War II a country that avoided major military involvements as persistently as European nations indulged in them, since then seems to have wholeheartedly embraced a doctrine of pre-emptive strike. No wonder Europeans practise a defensive policy and wait, with some apprehension, to see whether recent conciliatory US rhetoric will be synchronized with actual American *realpolitik*.

Notes

1 I. Mather, *A Brief History of the Warre with the Indians in New England* (Boston: John Foster, 1676), p. 1; quoted from R. Slotkin, *Regeneration through Violence: The Mythology of the American Frontier, 1600–1860* (Middleton, Conn.: Wesleyan University Press, 1973), p. 83.
2 George W. Bush, 'State of the Union Address', January 29, 2002.
3 George W. Bush, 'State of the Union Address', January 20, 2005.
4 R. Hofstadter, *The Paranoid Style in American Politics* (London: Cape, 1966).
5 Daniel Aaron, *The Unwritten War: American Writers and the Civil War* (New York: Alfred A. Knopf, 1973), pp. 140–5, sees the novel primarily as Twain's belated contribution to the Civil War, in which he did not serve personally. Given the comprehensive theme of the novel and its structural similarity to the 'captivity narratives', a more contextual socio-cultural interpretation seems appropriate.

6 See George Creel, *How We Advertised America* (New York: Harper & Brothers, 1920); James R. Mock and Cedric Larson, *Words That Won the War: The Story of the Committee on Public Information, 1917–1919* (Princeton, NJ: Princeton University Press, 1939); Harold Lasswell, *Propaganda Technique in the World War* (New York: Knopf, 1927); and George T. Blakey, *Historians on the Homefront: American Propagandists for the Great War* (Lexington: University Press of Kentucky, 1970). Susan Brownmiller, *Against Our Will: Men, Women and Rape* (New York: Simon and Schuster, 1975), p. 44, points to the far-reaching sexual aspects of the feminine allegorization of France in American propaganda of World War I.
7 W. Just, *Military Men* (New York: Knopf, 1970), p. 7.
8 W.J. Fulbright, *The Arrogance of Power* (New York: Vintage, 1966), p. 236.
9 J. Aho, *Religious Mythology and the Art of War* (London: Aldwych, 1981), p. 151.

SELECT BIBLIOGRAPHY

Aho, J., *Religious Mythology and the Art of War* (London: Aldwych, 1981).
Andrews, D. (ed.) *The Atlantic Alliance Under Stress: US–European Relations After Iraq* (New York: Cambridge University Press, 2005).
Ash, T. Garton, *Free World* (London: Penguin, 2005).
Bartlett, C.J., *The Special Relationship: A Political History of Anglo-American Relations since 1945* (London: Longman, 1992).
Baylis, J., *Anglo-American Defence Relations, 1939–1984: The Special Relationship* (London: Macmillan, 1984).
Baylis, J., *Anglo-American Relations since 1939: The Enduring Alliance* (Manchester: Manchester University Press, 1997).
Calleo, D.P., *Beyond American Hegemony* (New York: Basic Books, 1987).
Cooper, R., *The Breaking of Nations: Order and Chaos in the Twenty-First Century* (London: Atlantic Books, 2003).
Daalder, I. and Lindsay, J., *America Unbound* (Washington DC: Brookings Institution Press, 2003).
Dinan, D., *Europe Recast: A History of the European Union* (Boulder, CO: Lynne Reinner, 2004).
Dobson, A.P., *Anglo-American Relations in the Twentieth Century* (London: Routledge, 1995).
Dumbrell, J., *A Special Relationship: Anglo-American Relations in the Cold War and After* (London: Macmillan, 2001).
Dunn, D., *Poland – A New Power in Transatlantic Security* (London: Frank Cass, 2003).
Erb, S., *German Foreign Policy: Navigating a New Era* (Boulder, CO: Lynne Reinner, 2003).
Fulbright, W.J., *The Arrogance of Power* (New York: Vintage, 1966).
Grant, C., *Transatlantic Rift* (London: Centre for European Reform, 2003).
Hammond, B., *Banks and Politics in America* (Princeton: Princeton University Press, 1957).
Hill, C. and Smith, M. (eds) *The International Relations of the European Union* (Oxford: Oxford University Press, 2005).
Hunter, R.E., *The European Security and Defense Policy: NATO's Companion – or Competitor?* (Monterrey, CA: Rand Publications, 2002).

Kagan, R., *Paradise and Power: America and Europe in the New World Order* (London: Atlantic Books, 2003).

Kampfner, J., *Blair's Wars* (London: Simon and Schuster, 2003).

Kelley, J., *Ethnic Politics in Europe* (Princeton: Princeton University Press, 2003).

Lacoutre, J., *De Gaulle: The Ruler 1945–70* (London: Harvill, 1993).

Levy, D., Pensky, M. and Torpey, J. (eds) *Old Europe, New Europe, Core Europe: Transatlantic Relations after the Iraq War* (London: Verso, 2005).

Lind, M., *The Next American Nation: The new Nationalism and the Fourth American Revolution* (New York: The Free Press, 2005).

Lindstrom, G. (ed.) *Shift or Rift: Assessing US–EU relations after Iraq* (Paris: EU Institute for European Studies, 2004).

Longhurst, K., *Germany and the Use of Force* (Manchester: Manchester University Press, 2004).

Luttwack, E., *The Endangered American Dream* (New York: Simon and Schuster, 1993).

Marsh, S. and Mackenstein, H., *The International Relations of the European Union* (Harlow: Pearson, 2005).

Mellisen, J., *The Struggle for Nuclear Partnership: Britain, the United States and the Making of an Ambiguous Alliance, 1952–1959* (Groningen: Styx, 1993).

Merkl, P.H., *The Rift between America and Old Europe: The Distracted Eagle* (London: Routledge, 2005).

Meyer, C., *DC Confidential* (London: Weidenfeld & Nicolson, 2005).

Nathan, O. and Norden, H. (eds) *Einstein on Peace* (New York: Schocken Books, 1960).

Nicholas, H.G., *The United States and Britain* (Chicago: Chicago University Press, 1975).

Papacosma, S.V. and Heiss, M.A. (eds) *NATO in the Post-Cold War Era* (New York: St Martin's Press, 1995).

Pond, E., *Friendly Fire: The Near Death Experience of the Transatlantic Alliance* (Washington DC: EUS-Brookings Press, 2004).

Quinlan, J.P., *Drifting Apart or Growing Together? The Primacy of the Transatlantic Economy* (Washington DC: Center for Transatlantic Relations, 2003).

Rahv, P. (ed.) *Discovery of Europe: The Story of American Experience in the Old World* (Garden City, NY: Anchor Books, Doubleday, 1960).

Renwick, R., *Fighting with Allies: America and Britain in Peace and War* (London: Macmillan, 1996).

Richelson, J.T. and Ball, D., *The Ties That Bind: Intelligence Cooperation Between the UKUS Countries*, 2nd edn (Boston: Unwin Hyman, 1990).

Rupp, R.E., *NATO After 9/11: An Alliance in Decline* (New York: Palgrave, 2005).

Rynning, S., *NATO Renewed: The Power and Purpose of Transatlantic Cooperation* (London: Palgrave, 2005).

Seitz, R., *Over Here* (London: Weidenfeld & Nicolson, 1998).

Szabo, S., *Parting Ways: The Crisis in German–American Relations* (Washington DC: Brookings, 2004).

Vachudova, M., *Europe Undivided: Democracy, Leverage, and Integration After Communism* (New York: Oxford University Press, 2005).

Wood, G., *The Creation of the American Republic* (Chapel Hill, NC: University of North Carolina Press, 1969).
Yost, D.S., *NATO Transformed. The Alliance's New Roles in International Security* (Washington DC: US Institute of Peace Press, 1998).
Zabrowski, M., *Germany, Poland and Europe: Conflict, Cooperation and Europeanisation* (Manchester: Manchester University Press, 2004).

INDEX